EARTH'S FUTURE CLIMATE

by

Henry Willis

Llumina Press

Copyright 2002 Henry Willis.

All rights reserved. No part of this publication may be reproduced or transmitted in any form or by any means electronic or mechanical, including photocopy, recording, or any information storage and retrieval system, without permission in writing from both the copyright owner and the publisher.

Requests for permission to make copies of any part of this work should be mailed to Permissions Department, Llumina Press, P.O. Box 772246, Coral Springs, Florida 33077-2246, Telephone: 1-954-341-5636, Internet: www.llumina.com

ISBN: 1-932047-92-1.

Printed in the United States of America.

Library of Congress Cataloging-in-Publication Data

Willis, Henry, 1944-
  Earth's future climate / by Henry Willis.
      p. cm.
Includes bibliographical references.
  ISBN 1-932047-92-1 (alk. paper)
  1. Climatic changes--Environmental aspects. 2. Global warming. 3. Global environmental change. I. Title.
  QC981.8.C5W545 2003
  551.6--dc21

2003000963

Preface

The contents of this book are the result of a Meta review conducted on issues regarding global climate change. In that Meta review, I looked at over 6,000 articles on the Earth's climate and global climate. This book is in its essence a digestion of all those articles. I clearly wish to note the contribution of those others. There is an extensive bibliography that lists the source of all material used, and I wish to give credit to whom credit is due.

This book is for the person who wants to understand the complex issues surrounding global climate change, but who does not have an academic background to aid in that understanding. I wanted to level the playing field and provide a discussion on this subject that is not filled with scientific terminology that tends to be incomprehensible to the average person. To assist in that endeavor, there is an extensive glossary of terms. Further, much of the bibliography contains references to Internet websites. This will permit people without access to a university library to see what I saw when I conducted my research.

The other purpose of this book is to help people see through so much of the "hype" and personal agendas that currently surround any discussion of global climate change. I have tried to present a balanced discussion for both sides of the issue so the reader will have tools at his or her disposal to know what is true and what is not.

Do I have my biases? Yes I do, and I will tell you what they are right up front. I believe the evidence is strong enough to say with certainty that global warming is occurring. I believe that warming is occurring as a result of a periodic cycle in the variation of intensity in the output of the Sun's radiation. However, I also believe that the greenhouse gases produced by human beings are acting as an amplifier to increase the speed and intensity of this natural solar cycle. That may have unfortunate and unforeseen consequences in the return to another ice age in less than three to five years once that event begins.

Global climate change is the most complex question facing us today.

One person put it as "an interesting academic debate." Unfortunately, with the Earth's resources being stretched thin with an ever-increasing human population, we may not have the time or the luxury for an "interesting" academic debate. There is a need for as many people as possible to be knowledgeable about global climate change, so when the time comes, they have the ability to make an informed decision.

        Henry Willis
        Kailua, Hawaii
        January 2003
        E-Mail: henrypwillis@yahoo.com

Contents

Chapter 1: The Ghost of Our Creation	1
Chapter 2: Rapid Climate Change	5
Chapter 3: The Devil Is in the Detail	11
Chapter 4: The Greenhouse Effect	15
Chapter 5: Gas and More Gas	21
Chapter 6: Predicting Global Warming	27
Chapter 7: Mathematically Chaotic	33
Chapter 8: No Water, No Life	39
Chapter 9: Water, Water Everywhere	45
Chapter 10: The Masque of the Red Death	51
Chapter 11: How High Is High	59
Chapter 12: It's Not So Much What You Say	67
Chapter 13: Data Error	71
Chapter 14: Marine Worms and a Question	77
Chapter 15: Star Light, Star Bright	85
Chapter 16: Alternate Truths	89
Chapter 17: Time Machines	93
Chapter 18: Gyres	97
Chapter 19: It's Happened Before	101
Chapter 20: Ockham's Razor	107
Epilogue	113
Glossary	115
Bibliography	139

Chapter I
The Ghost of Our Creation

The question of global climate change is one of the most complex issues facing us today. It is well established that the Earth's climate has fluctuated through extreme cold to very warm periods and back again throughout the last 750 million years of its history. This fluctuation in the temperature of the Earth's climate is a naturally occurring event.

The Earth is entering into a period of its geological history where another ice age is possible at any time. Recent discoveries from ice cores drilled in Greenland found evidence that the change from a warm climate to an ice age climate occurred in a very short period of time, less than 3 to 5 years.

The change is coming. It will not be the change advocates of global warming predict. Rather, global warming will put into place a series of events that will cool the Earth's global climate to a point where it may bring on a new ice age. The irony is global warming can speed up the process that returns Earth's climate to an ice age. It began long ago.

There is a hiss at the outer ranges of the radio frequency spectrum. The hiss sounds insignificant like old A.M. radio background interference. Its wavelength of 1 mm on the radio frequency spectrum belies its apparent insignificance. The hiss literally stretched itself out to us across time and space. It is the ghost of our creation. It is all that is left of the "Big Bang," the explosion that created our universe 20 billion years ago. It started from that explosion as light with a wavelength of only a few microns. The light stretched to a lower frequency as it traveled through space and lost its initial energy. The farther the light traveled, the farther it stretched, until it stretched just far enough to be heard on the radio frequency spectrum.

This characteristic of stretching to a lower frequency of radiation is called the "red shift." Astronomers measure the distance to stars and how fast those stars are moving away or toward us by the "red shift". About thirty-five years ago radio technology advanced far enough to hear the hiss in the upper microwave radio frequencies. What we hear when we listen to the hiss are the last remnants of the light from the "Big Bang" that created our universe.

The "Big Bang" is only dimly understood. It began as a singularity. There was no time or space, and the present laws of physics did not apply. Then, that indescribable point exploded, creating all the heavens. It took time, lots of time, for the matter created by the "Big Bang" to organize itself. The first stage of organization was completed about 12 billion years ago. Stars of super dense hydrogen were born in which helium-forming fusion began. These stars were able to produce other elements up to the atomic weight of iron.

This continued until those stars ran out of fuel approximately 6 billion years ago. Then, they expanded into red giants, collapsed, and exploded into supernovas. Those explosions created all the other elements of the Universe.

Again, over time, organization of matter occurred, this time with the newly created elements. In our corner of the Universe, in an obscure arm of the Milky Way, that organization began about 4.5 to 5 billion years ago. This is when enough hydrogen and other matter gathered together for our Sun to start its nuclear engine.

Swirling in orbits around our Sun at this time were bodies of matter that became the planets and other stellar bodies of our solar system. In time, some planets almost became suns themselves. Some, too far from the Sun froze, while others too close to the Sun boiled. On only two of the planets did conditions occur to create environments that could support life. But Mars' gravity was too weak to hold an atmosphere or water vapor. Only on Earth were conditions right to allow the development of organic life.

Organic life first appeared on Earth approximately 2.5 billion years ago. This life survived in an atmosphere heavy with carbon dioxide. It continued to survive until the appearance of a genetic mutation—a blue-green algae that produced oxygen. The oxygen produced by the algae was a deadly poison to life dependent on carbon dioxide. The appearance of the blue-green algae 900 million years ago caused the first mass species destruction in Earth's history.

The appearance of oxygen in Earth's atmosphere was a fork in the evolutionary road. From this point on, the majority of life on Earth was dependent on the presence of oxygen. That life struggled against the severe climatic changes. Only in the last 5 to 10 years have scientists recognized just how unstable Earth's climate was 750 to 580 million years ago. During that period, the Earth completely froze over several times in events of total glaciations now called "Snowball Earth."

The Earth froze over during these events all the way from the poles to the equator; then it thawed out again through a sudden greenhouse effect. The period of time of total glaciations lasted approximately 10 million years. This killed most of the primitive organisms then living.

The "Snowball Earth" event occurred due to a lack of sufficient carbon dioxide in the atmosphere. As the Earth cooled, the carbon in living organisms was buried in the oceans, and there was a critical reduction in volcanic activity. This two-fold loss deprived the Earth's atmosphere of the necessary carbon dioxide to keep the planet warm. At some point as the polar ice caps expanded, global cooling accelerated because white ice reflects heat back into space more efficiently than the darker colors of land and open water.

The "Snowball Earth" did not last. Volcanoes returned to release carbon dioxide into the Earth's atmosphere. The levels of carbon dioxide in the Earth's atmosphere gradually built up to levels 350 times higher than those found in the Earth's atmosphere today. The melting ice released water vapor, the most powerful of the greenhouse gases, further accelerating the warming. Open, less reflective water returned. The melting of the ice that followed was extraordinarily rapid in geological terms.

Evidence suggests that the change from a "Snowball Earth" to a warm Earth and back again happened within a few hundred years. This cycle of a frozen Earth and extreme warmth repeated itself at least four times, possibly more, over a period of 170 million years. These events placed extreme stress on organic life.

On the one hand, organic life had to survive the extreme cold of a frozen Earth, and on the other hand, it had to adapt and survive the rapid change back to a warmer climate. At one point, life on Earth almost did not make it. The event was the Precambrian Extinction. Around 580 million years ago the Earth froze over again, and 95% of life on Earth became extinct.

Eventually, over a period of 30 million years, the Earth's atmosphere reached equilibrium in its content of oxygen and carbon dioxide. The

Earth's global temperature stabilized at 57° F, about what it is today. This was a favorable climate for the Precambrian Extinction survivors, and 500 million years ago those survivors caused an event called the "Cambrian Explosion."

The Cambrian Explosion was an unprecedented "explosion" (hence the name) of life. Almost all life we know today developed during the Cambrian Explosion. It was the largest expansion of life ever witnessed on Earth. The biological activities of Cambrian plants and animals guaranteed the stable equilibrium of the gases in the Earth's atmosphere. That equilibrium has lasted to present time.

Since the Cambrian Explosion, there were other species-ending events, the most well known of which is the disappearance of the dinosaurs at the end of the Cretaceous Period 60 million years ago. It is theorized that another major species-ending event occurred at the end of the Permian Period approximately 230 million years ago, when again approximately 95% of all life on the land and in the sea disappeared.

Six major and 23 minor species-ending events occurred over the last 500 million years. Many are concerned that mankind is creating its own species-ending event by the dumping of certain greenhouse gases into the Earth's atmosphere. For the first time since the advent of oxygen producing, blue-green algae, a species has the potential to alter life on Earth by changing the atmosphere. No one knows what that change will be. There are advocates who say the change will be a warmer global climate. In contrast, there is a smaller minority who believe if the Earth warms, for whatever reason, it will bring on another ice age.

Chapter 2
Rapid Climate Change

Every subject requires a form of introduction to construct a foundation of knowledge for what is to follow. As stated, the purpose of this book is to give the "average person" an understanding of the issues surrounding global climate change. The next several chapters are an introduction to that understanding. However, the scientifically-minded reader with knowledge of paleoclimatology may wish to skip this introduction and jump to Chapter 5 where the main discussion on the problems regarding global climate change begins.

There are problems anytime a subject as complex as global climate change is investigated. The Earth has a long history of global climate change. There is no question that global climate and its localized counterpart, weather, directly affect the quality of life. Those living in the United States endured two recent examples of this. The summer of 2000 saw a drought and the largest forest fire season ever in the United States. The fires raged out of control for most of the summer.

The winter of 2000-2001 proved itself to be one of the harshest on record for the Northern Hemisphere. It began with unseasonably early snowfalls. Average daily temperatures were 16° F lower on the average across the Midwestern and northeastern parts of the United States. There was unusually heavy rainfall across Europe. Worst of all, the temperature in Siberia and Mongolia dropped to an average daily temperature of -73° F, and the temperature stayed there for several months. The loss of life and property from the effects of these extreme winter temperatures in Siberia and Mongolia are still not counted. There is little doubt that the Earth's climate is changing.

Previous theory held that climate change is gradual. That is, the Earth's climate changes slowly over hundreds, if not thousands, of years. A quiet discovery little noted outside of scientific circles changed that theory. The discovery found that global climate does not change in the slow and stately manner as previously thought. The revelation was the discovery that a major global climate change can occur within just a few years.

The paleoclimate records (the record of climate before recorded history) indicate that approximately 11,500 years ago the Earth warmed to 9° F and 18° F in less than a decade. Then, during an event called the Younger-Dryas, it cooled right back down to ice age temperatures in less than 10 years and the temperature remained at ice age conditions for at least 400 years. The Earth since that event is now in an interglacial period (a warm period between ice ages) called the Holocene, where, with a few ups and downs, global climate has gradually warmed.

This discovery of rapid climate change is significant because of the impact such a change in Earth's global climate will have on modern civilizations. The discovery is compounded by current concerns about global warming due to the introduction of anthropogenic (human caused) greenhouse gases into the Earth's atmosphere. There is concern that humans are changing the chemical composition of the Earth's atmosphere with greenhouse gas emissions, creating the potential for an abrupt shift in global climate over a short period of time. The more rapid climate change is, the more damage the change causes. The social and ecological consequences of such a shift are incalculable.

Not a week goes by without some popular news source reporting a climatic or environmental catastrophe attributed to global warming. It is not difficult to understand why so many accept at face value the argument that greenhouse gases are causing global warming. It is immediate to anyone who visits a major industrialized city that there are environmental problems. The water is so polluted it must be heavily treated with chemicals. The air is so polluted with automobile exhaust, industrial smoke, and other vapor, visibility is sometimes less than a mile.

The summer temperatures in these cities almost always rise to 100° F, and higher. These are the reasons why people living in such an environment quickly grasp at the notion of global warming. They see in their everyday lives in what they believe to be the correlation between industrial gas emissions and a warming climate.

Care must be taken before coming to the conclusion that correlation is causation. I am old enough to remember the last global climate scare during the 1970s. The scare then was global cooling. Scientists monitoring global climate noted a small but significant downward trend in global temperatures. A good number of articles were written about the impact of decreasing global temperatures. They stated that industrial pollution was blocking sunlight from reaching the surface of the Earth, and this was the reason for the global cooling. This is the argument used today for global warming.

The error made during the 1970s was a baseline of measurement insufficient to draw an accurate conclusion. The conclusions were from a change over a short period of time compared to a change over a long period of time. It was a problem of resolution.

Resolution in the study of climate change is the accuracy of the measurement of time used. There are two ways resolution is noted. First, if the measurement of the time to when an event occurred can be dated to within a short period of time, the resolution is said to be high. The opposite is true if the measurement of the time to when an event occurred can be dated only within a long period of time. In that case, the resolution is said to be low.

For example, the time of a person's birth is usually known to the accuracy of a particular calendar date. This is high resolution. However, the date when humans first started using agriculture is imprecisely known to a period of time of several thousands of years. This is low resolution.

The second use of resolution is to express the length of time over which changing events are measured. Generally, when measuring change, the longer the period of time (that is, the longer the baseline) the more accurate the measurement. A long baseline of measurement is said to have high resolution, and a short baseline of measurement is said to have low resolution. A simple way to understand resolution is that high resolution is more accurate than low resolution.

The mistake made in the 1970s was a mistake of resolution based on observations of global climate over only a few years. The baseline was too short, and as a result, the resolution of the observations was low. Resolution was not the only error made. The second error was the assumption that there was a correlation between industrial pollutants in the atmosphere blocking sunlight and global cooling.

The same argument of correlation forms the basis for today's concerns regarding global warming. Simply stated, the argument is as follows: Humans are causing an increase in the release of greenhouse gas emissions into the Earth's atmosphere. Global climate is warming. Therefore, increased greenhouse gas emissions are causing global warming.

In scientific research, a correlation is an apparent relationship between two different variables (cause and effect). It is an established principle of scientific research that correlation is not necessarily causation. A correlation between two events is not proof that one event caused the other. We can only say that there is a co-variation between the two events.

A classic example of correlation error taught in many statistics classes is the apparent correlation between an increase in the murder rate and the increase in the sale of ice cream in Chicago during the summer. Every summer, the Chicago sale of ice cream and the murder rate increase. One might come to one of two conclusions: either the sale of ice cream increases the murder rate, or the murder rate increases the sale of ice cream. In fact, the probable cause for both might be the hot, humid, summer Chicago weather.

This is not meant to discredit the results of scientific research based on correlation. Statistical correlation can be a powerful scientific tool when properly used. A rigorously conducted study of correlation can provide significant insight into a scientific question, and sometimes a study of a potential correlation can statistically rise to the level of high probability.

The correlation between greenhouse gas emissions and global warming appears to be a logical statement on its face. However, the statement that greenhouse gases are causing global warming is a statement of a proposed correlation. This correlation has yet to pass the rigors of scientific investigation.

As with any statement of correlation, the relationship between the cause and the event is suspect. The cause, which in this case is stated to be greenhouse gases, may be something else, as with the Chicago ice cream-murder rate increase. The result, which here is said to be global warming, may, or may not, be true.

The clarity of the question and the likelihood of unbiased answers are further complicated by the ongoing media presentation of extreme worldwide climatic events. Instantaneous global communication is the marvel of our age. Little can happen in the world that cannot appear in our homes on the television, radio, or Internet within minutes. This means we can know about extreme weather event as they happen.

This is good because it informs those who are in harm's way. But it also sensitizes us; weather events that in another age were but a small article in the back of a newspaper are now reported live on television. This gives the impression of an increased frequency of serious weather events when the only thing happening is better communications about those events.

It is a great opportunity for the news reporting agencies. If news-reporting agencies can find a correlation between an extreme weather event and global warming, the event has even greater news worthiness. The curse is that these correlations are stated as scientific fact when just the opposite is true. The statements almost always go unquestioned and unchallenged.

This book questions and challenges the current and ongoing debate about global climate change. This is an effort to level the playing field so the average person can understand the complex issues. That is, I wish to let the farmer in Iowa, the rancher in Montana, the welder in Pittsburgh, and the stevedore in Los Angeles know in language they can understand what is happening with global climate. To that end, it is my position that it is not global warming that should concern us. It is global cooling.

Nature generally keeps things in balance, but Nature does not always maintain that balance in ways we expect. If the societal implications were not so compelling, the study of global climate change would be one of the most fascinating scientific problems of our time. Understanding the Earth's climate remains one of the major intellectual challenges faced by science today. However, global climate change also carries with it the potential for harm, suffering, and death, and it is one of the most serious challenges we face today.

The people in whom we put most of our trust to predict global climate change are scientists—for the most part well meaning people—unfortunately for the rest of us seem to speak in tongues. The science that should warn us about the challenges of global climate change is entangled in an apparent web of obscure and difficult to understand measurements and language.

An understanding of what the scientists are saying is critical to the understanding of the current discussion surrounding global climate change. This understanding lies within the tools scientists use. The tools are tools of measurement. Yet, even within the well-known measurements of temperature, weight and distance, and time, many differences exist.

Chapter 3
The Devil Is in the Details

Any discussion about global climate change will also involve a discussion of measurements. The three most common measurements encountered are: measurements of temperature; distance and weight; and time.

This is where the devil is in the details. Each of these areas of measurement has at least two different and distinct scales. Many writers on global warming switch back and forth between scales of measurement. If one is not familiar with both systems, this can be confusing. That is the devil; now on to the details.

Measurements of Temperature

There are three major scales in use today for the measurement of temperature. They are degrees Fahrenheit, degrees Celsius (formerly degrees Centigrade), and degrees Kelvin. Each scale is named for the man who invented it.

Daniel Gabriel Fahrenheit (1686-1736) invented the Fahrenheit Scale Thermometer. Fahrenheit's scale was initially calibrated to freezing water and the normal temperature of the human body. After Fahrenheit's death, scientists recalibrated the Fahrenheit Thermometer so 32° F is the freezing point of water and 212° F is the boiling point of water. The post-Fahrenheit version of the Fahrenheit Thermometer is now in limited use in some English-speaking countries, primarily the United States.

Andres Celsius (1701-1744) invented the Celsius Scale thermometer. The Celsius Thermometer is divided into 100 equal increments between the freezing point and the boiling point of water. What made the Celsius Thermometer more accurate than all previous thermometers is that Andres Celsius calibrated his thermometer at a specific barometric pressure (760 mm of mercury). The Celsius Thermometer is accepted as the standard for temperature measurements worldwide and within the scientific community.

There is one more temperature scale. It is degrees Kelvin, developed by William Thomson Kelvin (Lord Kelvin). Today, that for which Lord Kelvin is best remembered is his proposal of an absolute zero. Absolute zero is the theoretical temperature at which molecules of a substance have the lowest energy (no heat whatsoever). Many physical laws and formulas are simplified when an absolute temperature scale is used. This led to adoption of the Kelvin scale as the international standard for scientific temperature measurement in the 1970s. It is known as the International Standard (SI).

Measurement of Distance and Weight

There are two major systems in use today for the measurement of distance and weight. They are the Imperial System (sometimes called the "customary system") and the Metric System.

The customary system is a system of measurement that uses feet, pounds, and gallons. It comes down to us primarily from the Babylonians, Egyptians, Greeks, and Romans. The measurements are based on unrelated objects, phenomena, and human anatomy. The customary system of measurement eventually evolved into what is known as the Imperial System of Measures.

The problem with this system is that it lacks standardization among the countries using it. For example, the Imperial gallon used in the United Kingdom and the measurements of a gallon used in the United States are different, although both carry the designation of a gallon. This became a problem in the latter half of the 18th Century as commerce between nations increased.

The Metric System is a decimal system based on the length of the meter (from the Greek work, *metron*, to measure). Authorities give credit for the origination of the Metric System to Gabriel Mounton, a French vicar, at around 1670. Currently, the length of a meter is defined as the length of the path traveled by light in a vacuum during in $1/299,792,458^{th}$s of a second.

The beauty of the Metric System lies within its simplicity. All metric units are derived from the length of a meter. The unit of mass, the kilogram, is defined as "the mass of water contained by a cube whose sides are $1/10^{th}$ the unit in length" (a decimeter). The unit of volume, the liter, is defined in same way. Thus, one liter of water weighs one kilogram, and water weighing one kilogram equals one liter of water. If one knows the length of a meter, all other measurements of the Metric

System can then be derived.

The Metric System did not gain immediate popularity, and its adoption by the various countries of the world was slow. However, by 1975 every major industrialized nation in the world was using the Metric System with the exception of the United States. In this respect, it is somewhat ironic to note that in 1893 the United States Office of Weights and Measures (now known as the National Institute of Standards and Technology) officially adopted the Metric System in legally defining the yard and the pound.

Measurement of Time

Each scientific discipline has its own way of measuring time. That measure may be based on years ago, years before present, kilo years, carbon years, or a number of days ago. The way the passage of time is counted depends upon the scientific discipline of the writer.

Astronomers count the passage of time based on days ago from Earth's collision with an asteroid about 40,000 years ago. Social scientist use years before present (y.b.p.); historians use years ago; and physical scientists use carbon years ago (c.a.). Other disciplines use kilo years ago (k.a.), where one kilo year equals 1,000 years.

Operationalization of Definitions

Operationalization of definitions is the scientific way of saying, this is the way I have defined my terms. I recognize that most of my readers will be in North America. Therefore, temperature will be expressed in degrees Fahrenheit; weights will be expressed in pounds; distance will be expressed in miles; and the passage of time will be expressed in years before present (y.b.p.) from the year 2001.

This levels the measurement playing field. That is, I took all the different measurements I found, and I converted them to those systems. I did this for clarity. No reader will need spend time converting between systems of measurement, making the comparisons easier to see.

This clarification of the definitions of measurements is but a brief introduction into the complexity of the discussion about global climate change. As has also been stated here, the complexity is further increased under the considerations of resolution, correlation, and media hype. It does not get better. The further one delves into this question, the more complex it becomes. It starts with an understanding of what the greenhouse effect is, and why many argue it is influencing the Earth's future global climate.

Chapter 4
The Greenhouse Effect

The Greenhouse Effect is an expression often used in the discussion of global warming. The expression originated early in the history of orchid cultivation. Orchids are an ancient group of plants that evolved about 120 million years ago. Orchids were among the first of the flowering plants, and they can be found in every part of the world except Antarctica. There are approximately 35,000 different species of orchids in existence today, and the orchid is adaptable as to where it grows. Orchids can be found growing on trees, on the ground, on rocks, in bogs, and underground. Orchids are not parasitic. They grow on other plants, but they do not feed on those plants.

The popularity of orchids goes back before recorded time. Orchids are used as herbal medication in many places around the world. They are the symbol of love and beauty. The Chinese called orchids the "King's Fragrance." The Greeks thought they represented virility, and Medieval Europeans thought them to be an aphrodisiac.

The exotic fragrance and look of the orchid caused people to cultivate them simply for their beauty alone. Orchid cultivation in Europe was not established until the 1700s when a few orchids were brought to Europe by British sea captains. For years, botanists and wealthy amateurs were the only cultivators of orchids in Europe.

The beginning of modern orchid cultivation is credited to William Cattley. While unpacking a crate, William Cattley noticed the packing material was actually a strange, little plant. He potted some of the plants, and in November 1818, he successfully grew the first Cattleya Orchid, beginning the fascination of Victorian England with the orchid.

During the Victorian period, orchids were popular houseplants in England. However, they were rare because many orchids did not survive the long journey from the jungle to England, and orchids were difficult to cultivate in England's cold, wet climate. The popularity of orchids in

Victorian England led to the development of horticultural greenhouses made of glass. It was from the development of efficient Victorian greenhouses that we gained our basic understanding of the greenhouse effect of light.

How a greenhouse works is not difficult to understand. A greenhouse is made of glass. Glass lets light in, but it blocks heat from getting out. Light passes through the glass and strikes the ground of the greenhouse where it is absorbed. The visible light absorbed by the ground creates infrared radiation. We perceive infrared radiation as heat. The glass, which passed the visible light with no obstruction, blocks the passage of the infrared radiation and reflects it back toward the ground. As more visible light is passed through the glass, more infrared radiation is created that reflects back down into the greenhouse. This results in an increase of temperature within the greenhouse. A similar effect occurs in your automobile when it is left in sunlight. The heat felt inside the automobile is a result of the greenhouse effect.

Global warming advocates state this same effect occurs in the Earth's atmosphere. The Sun gives off a wide spectrum of energy, which we classify according to wavelength. Among the types of energy given off by the Sun are ultraviolet, visible and infrared light, also known as UV, visible and IR wavelengths of energy, respectively. Ultraviolet is a short wavelength, high-energy form of radiation. Visible light is a mid-wavelength, mid-energy form of radiation, and infrared is a long wavelength, low energy form of radiation.

The Earth's atmosphere is transparent to ultraviolet and visible light. When light from the Sun reaches the surface of the Earth as ultraviolet and visible light, the light is absorbed. This warms the surface of the Earth, and it causes the Earth to emit its own energy. Unlike the Sun, however, the Earth does not emit ultraviolet or visible light. It emits only infrared radiation. As this infrared radiation travels back toward space, it is reflected back to the surface of the Earth by certain gases in the atmosphere.

In this way, the Earth's atmosphere acts like the glass of a greenhouse, letting in the ultraviolet and visible light and absorbing and reflecting back infrared radiation. This warming of the Earth is called the "greenhouse effect."

Those gases in the Earth's atmosphere that block infrared radiation are called greenhouse gases. The most significant of these are carbon dioxide, methane and water vapor. They block infrared radiation from escaping the Earth's surface directly into space. Under normal conditions, most of the departing infrared radiation is carried away from the surface of the Earth by air currents and clouds, eventually escaping into space at altitudes above the thickest layers of the Earth's atmosphere.

The portion of the atmosphere that affects most of the Earth's climate is very thin, at best only about six miles thick. This is equivalent in thickness to the outer layer of the skin of an onion. In essence, greenhouse gases act like an insulator or blanket above the earth, keeping the heat in. It is argued that increasing the concentration of the greenhouse gases in the atmosphere increases the ability of the Earth's atmosphere to block the escape of infrared radiation. If the insulation of greenhouse gases becomes too thick, greenhouse gases will have a dramatic warming effect on the Earth's climate.

This will have repercussions, so that even slight increases in average temperature can result in big jumps in the number of warm days. If global warming puts more water vapor high into the atmosphere, the greenhouse effect will be even stronger.

Human activities are intensifying these natural greenhouse effects. Every year humans release billions of tons of heat-trapping gases into the atmosphere. In doing so, advocates for the greenhouse effect state humans are setting the stage for a warmer Earth.

Scientists keep track of the Earth's average temperature to understand global trends. To do this, they divide the Earth into a grid and collect temperature readings from all the weather stations in each grid. Averaging all of those measurements produces the figure that represents the average global temperature.

Whenever oil, coal, gas, or wood is burned, carbon dioxide is released into the atmosphere. Greenhouse advocates state that this and other gases we add to the atmosphere contribute to the global warming observed over the past 100 years, producing what scientists call a "discernible human influence on global climate."

Those arguing for global warming say this measurement of the Earth's average temperature revealed for the first time in Earth's history that humans are a decisive factor in future climate change. Human produced greenhouse gases are part of the reason for the 1° F rise in global average temperature documented over the last 100 years. The efficiency by which Earth radiates heat into space is reduced with a greater concentration of greenhouse gases.

This means that global temperature must rise in order for the same amount of heat to be radiated out into space. Earth's climate must adjust to rising greenhouse gas levels in order to keep its global energy budget in balance. In the long term, the Earth must get rid of energy at the same rate it receives energy from the Sun. Since a thicker blanket of greenhouse gases reduces energy loss, the Earth must adjust its climate in some manner to restore the balance between incoming and outgoing energy.

Global warming is a term that refers to the observation that the atmosphere near the Earth's surface is warming. This warming is one of many kinds of climate change that the Earth has gone through in the past and will continue to go through in the future. Global warming advocates believe there is evidence that climate change due to greenhouse warming began around 1900. Levels of all key greenhouse gases, with the possible exception of water vapor, are rising as a direct result of human activity. Emissions of carbon dioxide from burning coal, oil, and natural gas, emissions of methane and nitrous oxide, and emissions from agriculture and land use changes are changing the way the Earth's atmosphere absorbs energy.

Other emissions such as ozone from automobile exhausts and CFCs from industry are also contributing factors. This is all happening at an unprecedented speed, and it is known as the "Enhanced Greenhouse Effect." The Enhanced Greenhouse Effect is the name given to the effect of gases added to the Earth's atmosphere by human beings. (These gases are also known as "anthropogenic," or human-caused, gases).

Earth has a natural greenhouse effect that is necessary for life, as we know it. The Earth's climate is driven by a continuous flow of energy from the Sun that arrives mainly in the form of visible light. About 30% of this energy is immediately scattered back into space by Earth's atmosphere. The other 70% passes through the atmosphere to warm Earth's surface.

Water vapor and carbon dioxide found naturally in the atmosphere keep the Earth warm through the greenhouse effect. Without a naturally occurring greenhouse effect, the temperature of Earth would be about 60° F colder than it currently is. The greenhouse effect is important because it keeps the Earth at a habitable temperature.

However, problems may arise when Earth's atmospheric concentration of greenhouse gases is artificially increased. Many are worried that humans are altering the composition of Earth's atmosphere in such a way as to cause undesirable global warming. Scientists believe that greenhouse emissions from human sources are contributing to a worldwide warming trend. If this trend continues unchecked, the prediction is that the Earth's climate will warm by an average 2° to 6° F by the year 2100.

Carbon dioxide has an impact on global warming. A significant factor contributing to the potency of carbon dioxide's greenhouse effect is its persistence. As much as 40% of the carbon dioxide released into the Earth's atmosphere tends to remain there for centuries. The concentration of carbon dioxide levels for most of the last 10,000 years of Earth's history stood at around 270-289 ppm (parts per million). This is about a third less than the present concentration of carbon dioxide in the atmosphere.

Since the beginning of industrialization in the 19th Century, carbon dioxide concentrations in Earth's atmosphere climbed steadily. In the last thirty years, atmospheric concentrations of carbon dioxide increased sharply from 325 ppm in 1970 to 367 ppm in 1998. This accounts for nearly half of the total increase in atmospheric carbon dioxide concentration since pre-industrial times.

Chapter 5
Gas and More Gas

Atmospheric concentrations of carbon dioxide have increased by nearly 30% since the beginning of the Industrial Revolution. Energy burned to run automobiles and trucks, heat buildings, run power generation plants, and make cement is responsible for about 80% of all carbon dioxide emissions caused by humans. During the 1980s, humans released 5.5 billion tons of carbon dioxide into the atmosphere annually. Annual emissions of carbon dioxide during 1999 amounted to over 7 billion tons, or almost 1% of the total mass of carbon dioxide in the atmosphere.

Global warming advocates warn that if the use of fossil fuels continues to increase at present rates, by 2035, this human activity will contribute annually 12 billion tons of carbon dioxide into the Earth's atmosphere. This change in the atmospheric carbon dioxide level will increase the amount of energy absorbed by the Earth's surface by about 1.5 watts of energy per square yard. This is equal to about 1% of all the energy from sunlight that reaches the Earth's surface. It is argued by warming advocates that a substantial fraction of the excess carbon dioxide in the Earth's atmosphere will remain there for decades to centuries. About 15% to 30% of that carbon dioxide will remain in the atmosphere for thousands of years.

An important question in determining the influence of atmospheric carbon dioxide on future global warming is answering where does the carbon dioxide go? Human activity is currently releasing approximately 7.2 billion tons of carbon dioxide into the atmosphere per year. The Earth's oceans absorb approximately 2 billion tons of carbon dioxide per year. Soil absorption and plant photosynthesis account for another 1.6 billion tons. Another 2 billion tons are removed from the atmosphere by a yet unknown and unidentified source, called in scientific terms a "carbon sink."

This means according to those who argue for greenhouse warming that between 1.6 and 3.2 billion tons of carbon dioxide are accumulating in the Earth's atmosphere each year. It is an increase in the rate of carbon dioxide concentration of 1.5 ppm per year. This is a small but significant increase.

While carbon dioxide is a significant greenhouse gas, methane is another important greenhouse gas. Since the beginning of the Industrial Revolution, methane concentrations in the Earth's atmosphere more than doubled. Intensive agriculture, waste disposal, fossil fuel production, landfills, and coal mining are responsible for over 60% of the methane released into the Earth's atmosphere. The rest of the methane comes from natural sources such as wetlands and animal manure. (An interesting note is that almost two-thirds of the non-human contribution of methane to the Earth's atmosphere results from the activities of termites.)

The atmospheric concentration of methane increased from pre-industrial levels of approximately 775 ppb (parts per billion) in the 1850s to about 1,750 ppb today. From 1970 to 1999, the concentration of methane in the Earth's atmosphere increased by about 400 ppb. However, methane has an effective atmospheric lifetime of only about 12 years compared to the atmospheric lifetime of carbon dioxide that spans many centuries. Nonetheless, there is evidence that as the concentration of methane increases in the Earth's atmosphere, its atmospheric lifetime will also increase. This is important because it is estimated that methane has a 1,350% greater greenhouse effect influence than that of carbon dioxide. Global warming advocates estimate that the concentration of methane in Earth's atmosphere will double from the current amount of approximately 1,750 ppb to 3,600 ppb by the year 2100.

We have only begun to touch on the subject of greenhouse gases with the discussion of carbon dioxide and methane. It is argued that nitrous oxide, chlorofluorocarbons (CFCs), and ozone contribute the remaining 20% of the Enhanced Greenhouse Effect.

Nitrous oxide (also known as "laughing gas") is released by the use of nitrogen fertilizers, the burning of wood and some industrial processes. Nitrous oxide emissions rose by 15% over the last thirty years primarily due to more intensive agricultural activities. The atmospheric concentration of nitrous oxide increased from about 275 ppb in pre-industrial times to about 312 ppb in 1994.

Over the next 100 years the greenhouse impact of nitrous oxide may increase by 3 to 4 times. Those who argue for global warming project by 2100 the concentration of nitrous oxide in the atmosphere will increase from 391 to 433 ppb. This is an increase of 120 to 160 ppb over pre-industrial times.

A number of other greenhouse gases are also increasing in Earth's atmosphere. These include methyl chloroform, carbon tetrachloride, and numerous halocarbons. Ozone is both a naturally occurring greenhouse gas and an urban pollutant. Levels of ozone are rising in some regions in the lower atmosphere due to air pollution even as levels of ozone decline in the stratosphere.

What disturbs scientists the most about these inventories of greenhouse gases is a theoretical increase in energy striking the Earth's surface. The inventories suggest there will be an increase of about 1% of the net incoming solar energy that drives Earth's climate due to an increased concentration of greenhouse gases. This is equal to the amount of energy produced by burning 1.8 million tons of oil every minute, or said another way, it is over 100 times the world's current rate of commercial energy production.

The inventories of greenhouse gases scientists keep are called "emission inventories." An emission inventory is an accounting of the amount of air pollutants discharged into the atmosphere. The factors that characterize such an inventory are: the chemical or physical identity of the greenhouse gases, the geographic area covered, the life expectancy of the greenhouse gas in the atmosphere, and the kinds of activities that caused the emissions. A well-constructed emissions inventory should include documentation and data that allow readers to understand the underlying assumptions, not the least of which is a statement of which pollutants are being inventoried.

The primary greenhouse gases such as carbon dioxide and methane are considered to be stock pollutants. (Local air pollutants such as carbon monoxide, oxides of nitrogen and volatile organic compounds are not considered stock pollutants.) A stock air pollutant is a pollutant with a long lifetime in Earth's atmosphere that results in a greater atmospheric concentration over time. Stock air pollutants are also generally well mixed in the Earth's atmosphere. The impact of stock air pollutants is independent of where it was emitted. A greenhouse gas with this characteristic means it has a global influence.

The discussion thus far has focused on the human influence on the Earth's climate and the stock pollutants that are greenhouse gases. A second important human influence on climate is aerosols. Aerosols are very fine particles and droplets. The most consequential aerosols are sulfates.

Power plants that burn coal and oil; copper, lead and zinc smelters; burning of crop wastes; forest fires; and automobiles all release sulfur dioxide into Earth's atmosphere. Sulfur dioxide reacts with the water vapor in Earth's atmosphere to form sulfate aerosols. These aerosols are emitted in such massive quantities that they have a substantial impact on the Earth's climate (and the biosphere in the form of acid rain).

It is estimated that sulfur dioxide emissions worldwide will increase by 70% over the next twenty years. This projection is based on expected industrial growth in Third World nations as those nations begin to use more fossil fuels. While a 70% increase will not equate to a 70% accumulation (because aerosols are precipitated out of the atmosphere in a few weeks by rain), it is projected that sulfate aerosols will offset the warming effect of greenhouse gases by at least 7% over the next century.

There is one major greenhouse influence on the Earth's atmosphere that has so far been left out of this discussion. It is water vapor. The warming effect of carbon dioxide is small when compared to water vapor, even with the addition of human-caused, carbon dioxide emissions. Like carbon dioxide, water vapor stops the radiation of heat into space from the Earth's surface. Water vapor does this more efficiently than carbon dioxide.

The presence of water vapor in the Earth's atmosphere is not directly affected by human activity. Nevertheless, the influence of water vapor is a "positive feedback." In the discussion of global warming, a "positive feedback" is anything that increases global warming; conversely, a "negative feedback" (such as aerosols) is anything that decreases global warming. If other enhanced greenhouse gases cause a warming in the global climate, the atmosphere will retain more water vapor. It is predicted that even a little global warming will cause a rise of water vapor in the Earth's atmosphere.

This relationship between water vapor and other greenhouse gases is well recognized. As the Earth's climate warms due to human introduced greenhouse gases, the Earth attempts to cool itself by the evaporation of water. This puts more water vapor into the atmosphere where it stays because warm air holds more moisture than cold air. The presence of this additional water vapor in the Earth's atmosphere presents the potential for a further increase in global warming.

The effect of water vapor on global warming due to small increases in human introduced greenhouse gases into Earth's atmosphere is a major area of controversy. It is the existence and magnitude of water-related feedback that constitutes one of the greatest uncertainties in the predictions of global warming.

One of these uncertainties is the amount and type of cloud cover that will be produced by additional water vapor in the Earth's atmosphere. Certain kinds of clouds tend to have a cooling effect. But it is not yet known if those kinds of cloud cover will be produced by the additional water vapor.

The Earth will make adjustments. Warming is the simplest way for Earth's climate to rid itself of the extra energy. But even a small rise in global temperature will be accompanied by many other changes in, for example, cloud covers and wind patterns. Some of these changes may act to enhance global warming while others may act to counteract it. The problem lies in the ability to predict which is which.

Chapter 6
Predicting Global Warming

There is no end to man's fascination with the prediction of the future. It is a desire that existed long before written history. We find evidence of it in what remains of shamanistic religions throughout the world today.

A shaman is a person who claims an ability to contact a spirit world. It is through an ability to contact and control the spirit worlds that the shaman lays claim to the ability to cure diseases and see the future. There are many cultures worldwide today that still hold this belief.

The predictive methodology of humans is endless. It has in the past involved interpreting such things as animal entrails, dreams, omens from natural events, the coin toss of the I-Ching, and Tarot cards. Then, there are the people who are believed to have an ability to see the future.

The Bible records that Daniel was accurately able to predict the future through dreams. Both the ancient Greeks and Romans believed that the Priestess at the Temple of Delphi had the ability to predict the future. The ancient Romans believed that epileptics were able to see the future through visions. Julius Caesar was an epileptic, and his visions during a seizure were considered of great import.

This belief that special individuals have the ability to predict the future persists right up to modern times. Many continue to believe in the predictions of Nostradamus and more recently the predictions of Edgar Cayce. In this respect, many religions did, and still do, believe that certain persons can receive a divine revelation that will predict the future. Within Christianity, several such classical divine revelations are the Book of Revelations in the Bible, the visions of Joan D'Arc, and the predictions from Our Lady of Fatima. All major religions can provide similar examples.

Those who do future prediction hold their methodology is based within a (or *the*) paradigm of validity. What this means is there is a scheme or system to predict the future, if one only knows how to interpret it. All systems of prediction are based on this principle.

For example, both the ancient Chinese and the ancient Babylonians believed they found the paradigm in the stars and the planets. Both developed a form of astrology that uses the position of the stars and planets at one's birth to predict the future. It is a belief that continues to this day, as reported in the astrology columns of many daily newspapers.

Many scoff at these ancient efforts to predict the future. Yet, every one of the ancient systems of prediction was thought at the time of its use to be founded in a proven methodology. We of the 21st Century also have our magic to predict the future.

We call it science. Our science does not predict an exact future. It predicts the probability of what the future might be. It does so through two great inventions: mathematics and the empirical method of scientific investigation. We have gained such confidence in these two methods that we accept the conclusions made by them almost without question.

An understanding of what the future may bring is important for answering the questions surrounding global warming. There are presently two schools of thought on this. The first is, if more and better data regarding the critical components of the Earth's atmosphere are obtained, then better mechanisms of climate modeling can be developed to improve future predictions. The second school of thought is, because the Earth's atmospheric system is so complex, efforts to accurately model its behavior and predict future climate change is largely futile.

Whichever school of thought one chooses to accept, the balance of evidence suggests a human influence on global climate. More and more climate scientists are coming to the conclusion that human activity is causing the climate of Earth to change. The best estimate by those who argue for global warming is that about 50% of the observed global warming is now due to greenhouse gas increases emitted by human activity.

Compelling evidence demonstrates that global warming is already underway with consequences that must be faced. The evidence is of two kinds. The first are "harbingers." Harbingers are events that foreshadow the impact of events that are likely to become more frequent and widespread with continued global warming. Harbingers include spreading disease, earlier arrival of spring, shifts in plant and animal ranges, decline of certain plants and animals, coral reef bleaching, heavy rain, unseasonable snowfall and flooding, and droughts and fires.

It is hard to see when a harbinger of global warming is happening and this is a problem. Take a summer drought as an example. A drought begins when the rain stops. During the first few weeks, no one gives any serious consideration to the absence of rain. It is only when the lack of rain is felt through wilting crops, drying water reservoirs, and the like, that the drought is noticed. In the beginning no one saw it coming, and in the end everyone feels its influence.

The second are "fingerprints." Fingerprints of global warming are indicators of long-term global warming observed in the historical record. Fingerprints include heat waves, rise in sea level, coastal flooding, melting glaciers, warming Polar Regions, and increasing nighttime minimum temperatures. An analysis of past weather and climate ("fingerprints") is commonly used in the debate about global warming.

Care should be taken not to confuse weather with climate when using fingerprints to predict future climate change. Weather is the state of atmospheric conditions (hot and cold, wet and dry, calm and stormy, sunny and cloudy) that exists over relatively short periods of time from hours to a couple of days. Weather includes the passing of thunderstorms, hurricanes, blizzards, heat waves, or cold snaps. Weather's variability may be an unpredictable response to climate change.

Climate is the weather expected over a period of a month, a season, a decade, or a century. Climate is defined as "the weather conditions resulting from the mean state of the atmosphere-ocean-land system." In simple terms, weather deals with short-term events, while climate deals with the average weather conditions over a long period of time.

In terms of predicting Earth's climate warming through a review of known fingerprints, it is possible for a particular year to be the warmest year on record. The year 1998 was one such year—there was not a warmer year in the previous 100 years. The ten hottest years in the past century have all occurred since 1980, with 1995, 1997, and 1998 as the hottest years on record.

These changes are a cause for concern. They occurred in connection with the increased human emissions of greenhouse gases into the atmosphere. Greenhouse advocates argue the use of carbon fossil fuels caused a steady increase in both greenhouse gases and global temperature since the end of the last ice age approximately 12,000 years ago.

Under this prediction, the Earth's surface temperature is projected to increase by 1.8° F to 6.3° F over the next century if there is no abatement of carbon dioxide emissions. If carbon dioxide emissions continue to increase at present rates, a quadrupling of carbon dioxide will occur after 2100. In that case, projected temperature increases for such an atmospheric concentration are 15° F to 20° F above the present day temperatures.

The measurement of temperatures around the world over both land and sea began in 1860. Climate scientists combined all the separate readings and made estimates for temperature in places where data is sparse (e.g., deserts and mountains) in order to estimate the annual global temperature. These records show that the Earth is warming.

From 1970 to 2000, the average global temperature rose by 1° F. This is a speed of change that was unexpected. The change came from an unexpected source—not from an increase in the daily maximum temperature, but from an increase in the average daily lower minimum temperature. That is, the daily low temperature did not go down as far as expected.

There is one other indicator of future global warming that is both a fingerprint and a harbinger. It is the continued worldwide increase in human population. The current world population is estimated at 7 billion people. Assuming no significant increase in deaths due to warfare, disease, starvation, or some other disaster, by the year 2025 this number could be 14 billion, and by 2040 the population could reach 28 billion people.

This increase in worldwide population is considered by many to be a more serious threat in terms of global warming than all the industrial pollutants currently placed into the Earth's atmosphere by the developed industrial nations. All those people will have a need for energy. Currently, the burning of fossil, carbon-based fuels that release carbon dioxide and other greenhouse gases into the Earth's atmosphere provides that energy. This means that more people will release more greenhouse gases into the Earth's atmosphere.

From fingerprints and harbingers, scientists try to develop scenarios for the future. A scenario is a projection of what may occur in the future based upon currently known facts. It is from scenarios that computer modelers design programs to project future climate change. A scenario of any event must be viewed with care. All scenarios are based on assumptions, and those assumptions may be wrong. Therefore, it is important to know what the assumptions are for any scenario that predicts the future.

The scenarios for predicting future global warming loosely fall into three categories: the low scenario, the middle scenario, and the extreme (or high, or nonintervention) scenario. Each of these scenarios is based on the assumption that over the next century both world population and the world economy will grow. This will cause an increase in greenhouse gases released into the Earth's atmosphere, which will in turn increase global warming.

The extreme scenario (also called the high or nonintervention scenario) makes the assumption of nonintervention. Nonintervention means that no new policies are adopted to reduce the emission of greenhouse gases in response to the threat of global warming. This scenario assumes a world population of 11.3 billion people by the year 2100. It also assumes that the world's economy will grow by 3% per year over the next century. From these assumptions, this scenario projects that the annual worldwide emission of carbon dioxide will grow from 7 billion tons to 35.8 billion tons by 2100.

Most scenarios suggest that future growth in greenhouse gas emissions rates will be dominated by what happens in developing countries. The bulk of greenhouse gas emissions to date come from industrialized nations. Future increases in greenhouse gases are likely to come from emerging economies in which economic and population growth is the fastest. Projections about the industrialization and growth of emerging economies are the most uncertain of the projections made by global warming scenarios.

Chapter 7
Mathematically Chaotic

There are several factors that can affect the assumptions that all three scenarios use. Each scenario uses the implied assumption that there will be a sufficient availability of fossil fuels to create the increased greenhouse emissions. Many sources believe there are enough fossil fuel reserves to last another 300 years. Fossil fuels are a finite resource. What if the demand of developing nations for energy depletes those fossil fuel reserves not in 300 years but in 50 years? No fossil fuels, no greenhouse gases.

A second assumption that these three scenarios use is no development of alternative sources of energy. The 2001 energy crisis in California caused many homeowners to consider solar energy, and California and Colorado made, and are continuing to make, significant investments in the generation of electricity from wind power.

Solar panels for heating water in homes are already common in many places. The Japanese have on the market hybrid electric automobiles. Many automobiles in Brazil now run solely on alcohol produced by sugarcane. A power plant on the Big Island of Hawaii uses bagasse, which is a byproduct of processing sugarcane into sugar. India has developed a system to convert cow manure to methane for electrical generation in poor rural communities. (A byproduct of this system is a natural high-grade, fertilizer not made from petroleum.)

A number of cities in the United States are using the material collected from household trash to generate electricity. Forty percent of all products containing metal are recycled in the United States, reducing the demand for electricity to smelt metal oars. It is reasonable to assume that as the scarcity (and thereby the cost) of fossil fuels increases, alternative sources of energy will become more attractive and economically competitive.

Another assumption these scenarios use is a continued increase in world population. This assumption may be in error. Disease, warfare, and famine, and greater stress on limited resources, all possess the ability to reduce world population. There is a point at which the resources of the Earth cannot support a further increase in population. But no one knows what that point is.

The most serious flaw of predictive scenarios is the unforeseen event. Such events do occur, particularly when dealing with the unpredictable properties of the Earth's global climate. An event that we do not foresee is the most dangerous, and no scenario can assume such an event.

It is a "Catch-22." Scenarios are designed to predict. By the very nature of their design, they cannot foretell unpredictable events. Yet, to be accurate, unpredictable events must be factored in. If unpredictable events could be factored in, then, they would be predictable.

Many problems can arise from these basic assumptions of all scenarios. Yet, scenarios are important in a discussion of global climate change because computer programmers use scenarios to design simulations of global climate change. These simulations are called computer models, and computer models are used in an attempt to predict future global climate change.

Virtually all estimates of how the climate could change in the United States are the results of computer models of the Earth's atmosphere known as "general circulation models." These are complicated models. They are able to simulate many features of the climate, but they are still not accurate enough to provide reliable forecasts of how the climate may change. Further, different computer models yield contradictory results, and all computer models are dependent upon the accuracy of data fed to them.

Climate model projections fall broadly into two categories known as "Carbon Dioxide Doubling" and "Transient" Scenarios. The Carbon Dioxide Doubling Scenarios present an estimate of how the climate will change if the level of carbon dioxide in the Earth's atmosphere doubles within several decades.

These scenarios were common in the older versions of computer climate models. They generally analyzed how the Earth's global climate might change without attempting to calculate the influence of the oceans. They also do not consider the cooling effect of sulfates and other aerosols in the Earth's atmosphere.

Recently, elaborate computer models of the ocean currents were added to the computer climate models. These are the Transient Scenarios. Transient Scenarios use models that couple the ocean and the atmosphere into their calculations. Rather than simply calculating how the climate and oceans will respond to a doubling of carbon dioxide, these models use historic records to project changes in greenhouse gases in order to calculate how the climate might change year to year to some date in the future.

It appears that these ocean-atmosphere-coupled models are becoming more realistic in their projections. High performance computer programs simulate the important processes of the atmosphere and oceans, providing researchers insight into the links between human activities and major climatic events.

An example of this capability is the development of computer models to study the El Niño Southern Oscillation (ENSO) phenomena. These models are capable of predicting sea surface temperature anomalies in the tropical Pacific 6 to 12 months in advance. But, there are problems with their design.

Clouds present the greatest uncertainty about the extent of warming predicted by any given computer model. Computer models generally predict that cloudiness will change in a warmer world, when in reality the type and location of the clouds have varying effects. Clouds reflect sunlight, implying that more clouds will have a cooling effect. But many clouds, particularly those at high altitudes, also have an insulating effect. They are cold, and they are inefficient at radiating energy back into space. Thus, they help keep the Earth warm.

According to the computer models, the initial warming of the Earth's global temperature will be magnified by a series of positive feedback events (events that enhance global warming). The most important prediction is that surface warming will increase evaporation from the oceans and push more water vapor into the atmosphere. Water vapor is a strong greenhouse gas, and this will amplify the effect of the carbon dioxide in the Earth's atmosphere.

Again, the uncertainty is the effect that the cloud cover created by this additional water vapor will have on the Earth's atmosphere. Another uncertainty is how key sinks (processes that absorb or destroy greenhouse gases) will respond to a changing climate. For example, plants grow faster in higher carbon dioxide levels, and they absorb more carbon dioxide through photosynthesis (the carbon dioxide fertilization effect). Carbon dioxide fertilization, together with forest re-growth in northern countries, may absorb up to 25% of the carbon dioxide currently produced by human activity. No one knows how this sink will behave in the future, or if, as more land is required for food production, this trend will reverse itself.

The exchange of energy between the atmosphere and the oceans is another source of uncertainty for computer models. The speed and timing of global climate change strongly depends on how the Earth's oceans respond. The uppermost layers of the oceans interact with the atmosphere every year, and they are expected to warm along with the Earth's landmasses.

However, it takes over 40 times as much energy to warm the top 300 feet of the ocean as it does to warm the entire atmosphere of the Earth by the same amount. Ocean depths reach down several miles. This depth of the Earth's oceans will slow any atmospheric warming.

No one knows with certainty what cooling effect aerosols will have on global warming. Further, both the soil and ocean sediments have historically acted in the past as significant sinks for the absorption of carbon dioxide. No computer model of global climate change is currently able to factor in the influence of these carbon dioxide sinks on global climate change.

Also, no one understands the effect that ozone depletion will have on global climate change. The Earth's atmosphere is composed of several layers. We live in the Troposphere, where most of the Earth's weather occurs. Above the Troposphere is the Stratosphere. Ninety percent of atmospheric ozone is concentrated in the Stratosphere.

Normally, the oxygen that humans breathe has 2 oxygen atoms, and it is colorless and odorless. Ozone is a molecule containing three oxygen atoms. It is blue in color, and it has a strong odor similar to the burning of electrical wiring. Ozone molecules form a layer in the Earth's Stratosphere and are much less common than normal oxygen. Out of 10 million molecules of various gases in the Earth's atmosphere, about 2 million are normal oxygen, while only 3 molecules out of the 10 million are ozone. This very small amount of ozone in the Earth's atmosphere plays a crucial role in the absorption of radiation from the Sun, the most important of which is to protect the Earth's surface from the effects of ultraviolet radiation.

Loss of ozone in the lower Stratosphere was first observed in Antarctica in 1974 in what is now known as "Ozone Holes." It was eventually determined that ozone depletion in the lower Stratosphere was caused by a set of human-produced chemical compounds known as chlorofluorocarbons (CFCs). Once CFCs are in the atmosphere, intense ultraviolet radiation from the Sun cuts the chlorine molecules off. The unattached chlorine molecules convert ozone molecules into oxygen molecules, and this is what depletes the ozone layer.

Ozone is only a minor greenhouse gas. However, any depletion of ozone in the Stratosphere potentially has serious consequences for the health of plants, animals, and humans due to increased ultraviolet radiation. It was once thought that a reduction of ozone in the Stratosphere tends to reduce the greenhouse effect of other gases because the absence of ozone allows more heat to be radiated out into space. However, this may not be the case.

Ozone depletion allows more ultraviolet radiation to strike the surface of the Earth. Like visible light, ultraviolet radiation also generates heat when it strikes the surface of the Earth. This extra ultraviolet radiation may assist in the creation of the type of cloud formations that reflect heat back to the surface of the Earth rather than allowing the heat to radiate out into space. This is an effect not under current consideration by climate modelers.

By taking increases in greenhouse gases into account, computer models of the oceans and atmosphere can provide some general indications of what changes we might see in global climate in the future. Unfortunately, the capabilities of even the fastest computers, and our limited understanding of the links among various atmospheric, climatic, terrestrial, and oceanic phenomena, limit our ability to model important details of global climate change.

A climate suitable for human existence does not occur simply above some minimum threshold level of greenhouse gas concentration. Rather, it exists within a finite window, a limited range of greenhouse gas concentrations that makes life, as we know it possible. It may be that the Earth's climatic system is truly, in mathematical terms, a chaotic system not definable with precision. The only way to define it may lie in terms of general parameters of potential possibilities. It is, however, possible to predict with some certainty what will happen to our environment if global warming does occur.

Chapter 8
No Water, No Life

The Sahara Desert is an arid region of the world. This was not always so. Over the last 300,000 years this region alternated at least four times between a wet and a dry climate.

The Sahara of prehistory once had a vast network of rivers that are now buried by the sands of the desert. One of these buried rivers measured 12 miles across (as compared to the Mississippi River, which is 1 mile across at its widest), indicating it carried a large volume of water. This suggests a high level of rainfall at least during part of the year.

The last wet period for the Sahara occurred about 11,000 years ago. We know this from radiocarbon dating of ostrich shells and plant remains. At that time, the Sahara was a savanna-like environment of tall grasses, trees, and numerous lakes.

The alternation of northern Africa between a wet and an arid climate is important in the evolution of human beings and their eventual migration out of Africa. Recent genetic research into the origins of man indicate that humans evolved somewhere in southwest Africa around 300,000 years ago. We do know from archaeological evidence that humans were in South Africa around 110,000 years ago, and that they had migrated to Eurasia by 40,000 years ago.

By 11,000 years ago, human beings were in the Sahara. This is the time the last Ice Age was ending, and the Earth was entering the current period of general global warming in which we now live. This change in global climate caused the Sahara to collapse back into the arid state in which it is found today.

The return to arid conditions 5,000 years ago created a life-threatening situation for the human beings living in the Sahara region. The nomadic people of the eastern Sahara migrated to the Nile Valley to save their lives and what was left of their livestock. This did not occur without a struggle. There is archaeological evidence from gravesites in the Upper

Nile indicating warfare in the region between 4,000 and 5,000 years ago.

The warfare most likely occurred due to a dramatic increase in population in the Nile River Valley. Resources were scarce, and a struggle between people ensued to deal with decreasing and erratic rainfall and a river that flooded the land once a year. The need to feed an increasing population prompted the development of a complex society. That development occurred around 5,000 B.C. when Mena, also known as Narmet, became the ruler of the Upper Nile Valley and started the ancient Egyptian civilization of the pharaohs.

This is one example among many of how a change in the Earth's global climate influenced the course of human history. Today, there is concern in some scientific circles that human beings are causing a warming of the Earth. Of particular interest are the changes in regional climate and local weather, especially in the areas of extreme temperatures, heavy rainfall, and drought. Such events have staggering effects on societies, agriculture and the environment in the affected regions.

Global warming, if in fact it does occur, will bring changes in regional weather. Observations over land areas during the latter half of the 20th Century indicates that the minimum temperature has increased at a rate more than 50% greater than the increase of the maximum temperature. This increase in the minimum temperature lengthens the frost-free season. Frost in North America now ends 11 days earlier on average than it did in the 1950s. A longer frost-free season can be beneficial for many crops, but it will also enhance the growth and development of perennial plants and pests.

The reason why minimum daily temperatures are rising more rapidly than the maximum daily temperature is not currently known. One possible explanation is the increase in many areas of cloud cover and evaporative cooling. Clouds tend to keep the days cooler by reflecting sunlight back to space and the nights warmer by inhibiting loss of heat from the Earth's surface. Further, greater amounts of moisture in the soil from additional precipitation and cloudiness inhibit daytime temperature increases because part of the solar energy is used to evaporate this moisture.

Energy from the Sun's evaporating moisture is what works the Earth's global hydraulic cycle. The heat from the Sun evaporates the moisture from bodies of water and soils. The evaporated water forms into a mist we see as clouds. Clouds are water vapor, and the water vapor maintains itself as a cloud for as long as it retains the heat it absorbed from the Sun.

The occurrence of rain is largely determined by relative humidity. Humidity is the ratio of the concentration of water vapor to its maximum saturation value at a given temperature. When the relative humidity reaches 100%, the water in clouds condenses into rain. Generally, this maximum saturation occurs when a cloud is cooled, when it encounters particulates in the atmosphere such as sulfate aerosols, or when it encounters an obstacle such as a mountain.

Global warming is expected to increase evaporation in the Earth's hydraulic cycle. As happened in the Sahara Desert when the last Ice Age ended, global climate change can affect rainfall and change the global patterns of drought and floods. Rainfall will not increase everywhere, and it will not occur throughout the year. Rather, the distribution of rainfall is dependent upon atmospheric circulation that transports moisture. Said another way, where and when it rains depends largely on global wind patterns.

Rain carried to a particular place over a long distance by global wind patterns is different from rain generated by local evaporation (evaporation from water and soil near where the rain occurs). The Asiatic Monsoon is a classic example of rain carried over long distances by global wind patterns. If global warming changes the path of the Asiatic Monsoon, many areas dependent upon those monsoon rains may not receive this needed moisture.

On the other hand, local evaporation is dependent on available water in the soil to create rain in the area. In a warmer climate, there will be increased evaporation from the soil in the spring. This will dry out the soil, leaving less water available in the soil for evaporation and rainfall in the summer.

There is a general consensus that annual worldwide precipitation and evaporation will increase a few percentage points for every degree of warming. Many scientists believe that the middle latitudes of the United States will see drier summers due to this evaporation effect. In fact, some are of the opinion that the extreme and extended drought in Texas during the summer of 2000 was caused by this effect.

On a larger scale, it is predicted that global warming will result in increased average precipitation in winter at high latitudes. High latitudes are the degrees of latitude that are closer to the Earth's Polar Regions (conversely, low latitudes are the degrees of latitude closer to the Earth's equator). Heavy precipitation will move from the tropics to higher latitudes. There will be less rain in the tropics and more rain in the Polar Regions.

This change in precipitation patterns is happening. Since 1900, precipitation has increased during the winter in the high latitudes of the Northern Hemisphere. But, for tropical and subtropical land areas, precipitation has decreased over the last several decades. This is especially apparent over the Sahel in Africa, as well as Indonesia, and New Guinea. The last two of these countries lie in the path of the Asiatic Monsoon winds, and a reduction of annual precipitation in these areas may be an indicator of a shift in the global pattern of the winds that bear the Asiatic Monsoon.

It is the opposite side of the precipitation coin that most concerns scientists, and that is the increased frequency of heavy downpours of rain. The concentration of the water vapor in clouds needed to reach saturation (the point at which it rains) increases with the temperature of the air at about 6% for every 1.25° F. Clouds will carry more moisture in a warmer climate, and that will increase the events of extreme rainfall.

The frequency of extreme rainfall increased throughout most of the United States over the last 100 years by about 5%. Along the northern tier of the United States and southern Canada, there was an even higher increase of such events by about 10% to 15% over the last 20 years. Most of this increase is taking place between September and November, exactly the time of year for heavy rainfall predicted by global warming.

Rainfall is also tending towards heavy, concentrated downpours throughout all of the United States. At the beginning of the 20th Century, only 9% of the United States experienced a storm in which more than 2 inches of precipitation fell in a 24-hour period of time. In the last decades of the 20th Century, such a severe storm occurred each year over 11% of the United States. The number of extreme events of rainfall (more than 2 inches in 24 hours) increased in the United States by more than 20% since 1900.

These indicators conspicuously stand out in the picture of surface climate variations and changes in the United States over the past 100 years. The events include a steady increase in rainfall derived from extreme one-day precipitation events, the continued decrease in the day-to-day variations of temperature, and the increased frequency of days with some form of precipitation. This accelerated hydrologic cycle means there will be greater precipitation in some areas; other areas a greater frequency of droughts in the summer; and floods in the spring and fall throughout the United States.

The water resources of the United States are sensitive to both rising temperatures and changes in precipitation. Water will become one of the key concerns of the United States in upcoming years. Droughts, floods, declining snow packs, water quality, and water use are all issues representing potential conflicts.

The increase in the global daily minimum temperature and the change in global rain patterns will impact other areas where water is involved. The foremost of these secondary impacts lies in the melting of the Earth's ice. In the next 100 years it is predicted between 1/3 and 1/2 of the world's mountain glaciers will melt. This will affect the water those glaciers supply to rivers, hydroelectric dams, and agricultural irrigation.

Glaciers are present on every continent except Australia, and they are excellent indicators of global climate change. There is clear evidence that the Earth's valley glaciers, ice caps, ice fields, and associated outlet glaciers are in recession. "Recession" is the geological term for a glacier in retreat, or said in another way, a glacier that is shrinking.

Earth's ice caps, non-tidewater and non-surge-type glaciers have been in recession since the early 19th Century. Observations of mountain glaciers situated on the equator in both Africa and South America indicate that the speed of glacial recession is increasing. A recent survey of Swiss Alpine glaciations found that those glaciers retreated by an average 300 feet per year between 1850 and 1973.

Further, no one knows with any certainty what is happening to the Earth's polar ice. The problem is that both of the Polar Regions of the Earth are so isolated and hostile that it is difficult to get continuous climate measurements from those regions. It is known the Arctic Sea ice has decreased since 1973 by about 14,000 square miles per year. The West Antarctic Ice Sheet is also vulnerable to global warming because its geological makeup allows it to be undercut by warming ocean water. The problem is, it is unknown if these events are indicators of global warming or simply events of a larger and ongoing expression of climate variability.

Chapter 9
Water, Water Everywhere

The Earth's mountain glaciers are melting, and it is assumed from this that polar ice situated in the Arctic Sea, Greenland and Antarctic are also melting. However, the increased precipitation from global warming may be hindering the melting of the Earth's polar ice through increased precipitation that will result in more snow falling in the Polar Regions. That snow, in turn, will turn into ice. This increased production of ice may counterbalance the increased melting. This is precisely what is happening with the Patagonian Glaciers in South America.

The loss of the Earth's ice cover from glaciers and sea ice can amplify the effects of global warming. Snow and ice efficiently reflect sunlight while bare ground and open water absorb heat. If a small amount of warming melts snow earlier in the year, more energy from the Sun will be absorbed by the exposed ground and water, causing more warming.

A corollary to the melting of the Earth's glaciers and sea ice is a rise in sea levels. Some scientists are of the opinion the melt water from the world's glaciers contributed to a rise in sea level during the last century. Worldwide measurements from tidal gauges during the last 100 years indicate that the global mean sea level rose between 6 and 8 inches. This increase is greater than can be explained from the geological record over the last 2,000 years, if it is true.

Approximately 1 to 2 inches of the increase in the global mean sea level is attributed to melt water from glaciers and sea ice. The remaining 4 to 6 inches in increase is attributed to the thermal expansion of the upper layers of the oceans as they warm. When water is heated, it expands. The water of the oceans is no exception. When the upper layers of the water in the oceans are heated by an increase in air temperature, those waters expand. The expanding ocean water increases the global mean sea level.

The global mean sea level of Earth's oceans is predicted to rise about

12 inches by the year 2100. However, the range of uncertainty in this prediction is great between 6 and 24 inches. There are several factors for the uncertainty.

First, a significant amount of sea level increase is due to the expansion of ocean water from a warming climate. Second, land areas covered by glaciers during the last Ice Age advance are rising due to the removal of the weight of the ice. The sea level is dropping relative to those coasts. Last, no one currently knows what effect on sea level the melting of the great glaciers of Greenland and the Antarctica will have.

The vulnerability of the West Antarctic Ice Sheet is poorly understood. Some scientists believe it will slide into the ocean after a sustained warming of global temperature. The West Antarctic Ice Sheet contains enough ice to raise the Earth's sea level by 20 feet. There is geological evidence to suggest that such a melting occurred approximately 110,000 years ago. If this is accompanied with the melting of the Earth's mountain glaciers, the rise in sea level could be as much as 30 feet. This will swamp many oceanic islands and redraw the world's coastlines.

Any increase in sea level also brings an increased risk of storm surges. Currently, there are 46 million people around the world at risk of flooding from storm surges. A 1-1/2 foot rise in sea level will result in a risk to 92 million people. A 3-foot rise in sea level will increase that risk to 118 million people. Further, a 6-foot increase in sea level will be enough to flood 1% of Egypt, 6% of the Netherlands, 17.5% of Bangladesh, and 89% of Majuro Atoll in the Marshall Islands. A 30-foot increase in sea level will destroy almost every coastal city in the world.

(However, as discussed further on, the stated increase in sea level from global warming is yet to be proven. There is considerable evidence to suggest it is not happening. Also, there are a number of variables that must be calibrated before any statement about mean sea level can be made.)

Coastal flooding notwithstanding, if there is a severe weather event (outside of a tornado) that most concerns people in the United States, it is a hurricane. The intensity of a hurricane is a result of its interaction with the ocean. The warm water at the top of the ocean provides the heat that fuels a hurricane. As it spins, the hurricane churns the water beneath it, bringing up cooler water from the ocean's depths. The colder the water, the less fuel the hurricane has. In the Caribbean, where even the deep water is warm, hurricanes tend to be the most powerful.

The intensity of Atlantic hurricanes has varied with the intensity of the surface temperatures of the Atlantic Ocean over the last century. The number of hurricanes increases when the ocean water is warmer. As the climate warms, scientists anticipate hurricanes (also known as typhoons and tropical cyclones) will vary by geographic region. Not all the consequences of hurricanes will be negative. In some areas, the contribution of tropical cyclones to rainfall is crucial. In northwest Australia, for example, 20% to 50% of the annual rainfall is associated with tropical cyclones.

Early discussions of the impact of global warming on hurricanes often suggested more frequent and intense storms. The argument went something like this: These storms depend on a warm ocean surface with an unlimited supply of moisture, forming only over oceans with a surface temperature of at least 79° F. Global warming will lead to an increase in ocean temperatures, which in turn will cause increased hurricane activity.

This, to date, has not proven to be the case. Reliable records of the number and intensity of hurricanes that reach the United States dates back to 1900. A study of hurricane intensity indicates that the frequency of hurricanes that make landfall on the United States was relatively low over the last decades of the 20th Century when compared to hurricanes early in the century.

On the other hand, it is expected that global warming will leach the moisture from the soil during the summer months. This will cause extended periods of drought. Then, the moisture stored in the atmosphere will be released in the fall and winter months in events of extreme rainfall and snow.

This prediction appears to have already rung true for the United States. In the early 1990s, two 100-year floods occurred in less than five years in the Midwestern United States. The summer of 2000 saw a severe drought in Texas, Florida, California, and the northwest United States that resulted in widespread, uncontrollable forest fires. Increased precipitation will fall mostly as rain rather than snow in a warmer global climate.

Less water will be stored in snow packs because the water will run off immediately, adding to winter flooding and landslides. An early snowmelt that increases spring flooding also decreases the availability of water from the snow in the summer. The need to ensure summer and drought-condition water supplies will lead water managers to keep reservoir levels higher. This will limit the capacity for additional water retention during unexpected periods of unusually heavy precipitation.

The precipitation that does occur will be concentrated in a few heavy storms rather than many light showers. This will shift flood plane boundaries. Most flood-prone communities of the United States are at least partly protected by levees and flood storage reservoirs. However, as the Mississippi and Missouri River floods of 1993 illustrated, the flood protection systems in the United States are designed to provide protection from only the moderately destructive floods.

A moderately destructive flood is defined as one with a 1% chance of occurring in any given year. Flood protection systems designed to protect at this level are ineffective against those rare floods at, or greater than, the level of the 1993 floods in the United States. Increased flooding does not increase the amount of water available for human consumption, absent efficient mechanisms for the capture of that floodwater.

The most common system for the capture of floodwater is a dam. However, there are only so many locations in the United States where dams can be built, and most of those locations already have a dam. For example, there is no place along the lower Mississippi where a major flood control dam can be built. The land is too flat. The building of such a dam would require the flooding of millions of acres of valuable farmland for its reservoir.

Floods have other indirect impacts on water available for human consumption. Floodwaters tend not to trickle down and replenish aquifers. Floods overwhelm sewage disposal systems and pollute otherwise fresh water. Floods do not increase the annual average volume of water flow in rivers.

Floods are a specific event unrelated to the average volume of water that flows through a river. Once the flood has passed, the river returns to its normal level and rate of flow. The predicted impact of global warming on precipitation patterns will increase the magnitude of floods on the rivers of the United States.

This predicted impact also assumes a smaller amount of snow in the mountains and little or no rain during the summer months. This means there will be a large volume of water in the rivers for a short period of time, and then the level of the rivers will drop due to the lack of rain and snowmelt. Decreased river levels will have a direct impact on the amount of hydroelectric power generated. The amount of hydroelectric power generated depends on the flow of water volume.

A 1% decrease in the flow of water volume produces a greater than

1% decrease in hydroelectric power production. Not only does less water run through the turbines that produce the electricity, but also, lower reservoir levels reduce the water pressure. Reduced water pressure reduces the power produced by a given volume of water flow. For example, a 10% decrease in the volume of water flow in the Colorado River's lower basin results in a corresponding 36% reduction in the production of electricity.

Many of the major rivers in the United States are used for barge navigation. The three most significant rivers in the United States used for this purpose are the Mississippi, Missouri, and Ohio Rivers. The United States Corps of Engineers built an elaborate system of locks and dams to eliminate shoals and rapids in order to maintain water flow for navigation in these rivers.

These structures are affected by erratic fluctuations in runoff from precipitation variations. A change in the year-to-year variability of climate due to global warming, such as severe floods or too little water, threatens navigation on those rivers. The 1993 Mississippi and Missouri River floods demonstrated that a wetter climate can overwhelm those rivers' navigational infrastructures.

During the 1993 floods, navigation on those rivers was suspended at a flood stage level considerably lower than the final level of the flood. Another concern for navigation in the lower Mississippi River is that rising sea levels will directly impair the ability of the Corps of Engineers to maintain the shipping lanes south of New Orleans.

There will be both a positive and negative impact on Great Lake shipping from global warming. There is an expected decline in the levels of the Great Lakes due to evaporation from global warming. A reduction in the levels of the Great Lakes will reduce the flow and depth of the St. Lawrence River. A reduction of the St. Lawrence River's flow and depth will reduce the tonnage that ships can carry, and it is estimated that for every 2-inch drop in the St. Lawrence River, the cost of shipping on that river will go up by 1%. On the other hand, warmer temperatures will extend the ice-free days on Lake Erie by as much as three months.

Another impact on the rivers in the United States is the quality of the water flowing in those rivers. A decrease in the volume of water flow will increase the salinity of the water. It is currently estimated, for example, that the salinity of the Colorado River will rise by 15% to 20%, not only affecting the availability of fresh water in the western United States but also in northern Mexico.

Further, the decreased depth of rivers from a reduction of flow will result in an increased temperature of the river's water. Higher water temperature will decrease the solubility of oxygen in the water. This will kill fish and other aquatic life.

Warmer water hastens the rate at which organic pollutants degrade. This degradation exerts a "biochemical oxygen demand" (BOD). BOD further decreases the amount of available oxygen in water. It permits the growth of certain kinds of algae and pathogens that infect the wildlife dependent on the water. Infected wildlife can in turn infect the human population. BOD also limits the ability of municipalities to use the water for sewage disposal, creating health problems from improperly disposed waste.

In the discussion of the impacts that will result from global warming, water is another case of a curse and a blessing. Water vapor is by far the strongest of all greenhouse gases. Yet without water life as we know it is not possible. Water vapor may form clouds that will protect us from an increase in global climate temperature. On the other hand, the increased amount of water vapor in those clouds may also lead to a change in global precipitation patterns that influence water resources in a way not currently known.

The availability of fresh potable water is a boon. But too much or too little of it creates problems. Global warming is projected to have an impact on the availability of fresh water through changed precipitation patterns.

Chapter 10
The Masque of the Red Death

There is a connection between mummies found in the Tarim Basin of northwestern China and global warming. The connection lies within the group of languages to which English belongs. The migration of the people who used those languages put in place a set of events that led directly to the Industrial Age and the release of greenhouse gases.

The English language is a member of a subgroup of languages called Teutonic. In turn, the Teutonic subgroup of languages is a member of a larger family of languages called Indo-European. The British jurist Sir William Jones discovered in 1786 that Latin and Greek shared a common origin with Sanskrit, the ancient language of Hindu law and religion. He proposed that all three of these languages developed from a single parent language now called Proto-Indo-European, or "PIE" for short. This set off one of the most contentious and longest-standing debates in linguistics (the science that studies languages).

The source of the debate is the broad diversity of the people using languages belonging to the PIE group. Linguists discovered that languages belonging to the PIE group are used all the way from Scotland to the Buddhist monasteries and caravan cities of the Tarim Basin in western China. The two questions that arose were: How did the PIE group get to such diverse localities? And, how long ago did it occur?

We know that the last Ice Age reached its maximum area of coverage approximately 22,000 years ago. The glaciers blocked human migration into northern Europe and Asia. However, approximately 12,000 years ago, the Earth began to warm, and the Ice Age glaciers went into a rapid retreat. Prior to the retreat of the last glaciers, 40,000 to 70,000 ago, prehistoric man came out of Africa to the Middle East, India, Southeast Asia and Australia.

Man's early ancestors also migrated into southern Europe, where they were in part responsible for the demise of Neanderthal man around 22,000 years ago. Then, 10,000 to 12,000 years ago, as the northern ice began to melt, humans migrated into the land newly opened by the retreating glaciers.

It was long thought that this migration into Asia was from the south through India and Indochina. It is true that migration of man into Asia did occur along this route. There was, however, another route, revealed when the Tarim Basin mummies in western China were discovered. The mummies (not really mummies, but bodies preserved by a very dry climate) date to between 1800 B.C. to 2000 B.C. What made the discovery significant is that the mummies had Caucasian features in an area of the world populated today by people with predominately Asiatic features. This is where the anthropological disciplines of archaeology and linguistics came together.

It is unlikely the people who originally spoke the PIE languages came north from China where an entirely different language group exists. It was demonstrated that the weaving patterns of the cloth found on the mummies in the Tarim Basin matched weaving patterns of cloth found in Scotland from the same period of time. Also found buried with the mummies were horses with saddles and bits. It was the horse bit that proved to be the clue that solved the mystery of how Caucasians came to western China.

Analysis of the horse bits and bite marks made by the horses on the bits revealed that the people who buried the Tarim Basin mummies rode their horses in the same manner as the people who inhabited the Volga River region of southern Russia between 5000 B.C. and 4500 B.C. The people of the Volga River region had domesticated cattle and sheep, and they were possibly among the first humans to use the domesticated horse for transportation.

The people of the Volga River region migrated eastward over the years into what is now modern Afghanistan, then to the Tarim Basin region around 2000 B.C., right at the time of the beginnings of the formation of the Chinese civilization. It is highly likely that the Tarim Basin people played an important role in the east-west diffusion of technology and trade, and they were responsible for starting the first transcontinental exchange of trade goods, and ideas.

Three thousand five hundred years later, the transcontinental, migration route established by the Tarim Basin people was institutionalized as the Silk Road. The Silk Road was for centuries the main route of trade between western China and the Middle East. Trade routes, historically, also acted as routes for invaders. In the early 14th Century, one such invader left western Mongolia and traveled the Silk Road westward to the Middle East.

The invader was the *Pasteurella pestis* bacterium, also known as the Black Death. It appeared in a ship off Alexandria, Egypt, in April of 1325. Bubonic Plague was not unknown to this part of the world; the first record of the plague in Europe occurred in Athens, Greece, in 430 B.C. One of the worst appearances of the Bubonic Plague in the ancient Mediterranean world occurred in Rome in 262 A.D. The chroniclers recorded 5,000 people a day died for half a year. If these records are correct, that outbreak killed over 900,000 people.

The outbreak of the Bubonic Plague in Egypt in the early 14th Century was no less devastating. Returning Crusaders carried it to Western Europe, and by the time it ran its course, 45% to 60% of the populations of Europe, 60 million people were dead. Bubonic Plague is a devastating disease when it occurs in a population unprotected by modern medicine. But the mortality rate from Bubonic Plague in Western Europe in the 14th Century was significantly higher than what might be expected.

One theory is that the malnutrition of Western Europe's population made it more susceptible to the impact of Bubonic Plague. Historical records reveal that many people in Europe were indeed malnourished. (However, a recent discovery of medieval graves in Scotland suggests that anthrax "piggybacked" with the Bubonic Plague, increasing the mortality rate.)

One must step back a bit in time to understand the malnourishment of Western Europe during the 14th Century. Between 1000 and 1300, the climate of Western Europe warmed in an event known as the Medieval Warming Period. The climate during this period in Western Europe was so mild that wine grapes were grown in Scotland. Agricultural production improved, and the improved agricultural production supported an increased population.

Sometime shortly after 1300, the climate in Western Europe abruptly cooled in an event now called the Little Ice Age. The temperatures dropped on average from 5° F to 10° F. What had been a mild, wet climate became a cold, wet climate. Agricultural production dropped. Surviving historical records indicate that this brought on mass malnutrition and starvation among the feudal serfs. Then the Bubonic Plague struck, and the resulting loss of life from the Bubonic Plague caused a breakdown of social structures.

There was societal chaos (including major warfare). Labor became scarce and valuable. The response to labor shortages was the Industrial Revolution, which began in Great Britain in the early 18th Century with the invention of steam power. Industry required power and that power was first derived from coal, and later on, from petroleum. Each of these fossil fuels produces the greenhouse gases that are currently under scrutiny for global warming.

The foremost threat from global warming to the Earth's increased human population is disease. There are already on hand a number of diseases that have the potential to adversely affect the human population. AIDS is not under control. It is only a matter of time before Ebola escapes the African continent, and the ancient killers of malaria and tuberculosis are both on the increase, with new varieties resistant to modern medicines.

No one under the age of 40 is immune to small pox if that scourge should ever again be released on the world. The mosquito carries serious diseases that can cause death in human beings, and the mosquito is becoming resistant to the pesticides designed to control it. The flu virus in mutated form can kill as it did during the 1918 influenza epidemic when 18 million people died. And there is the ever-present threat from Bubonic Plague, which continues to show up from time to time throughout the world.

One result of global warming is that the frequency of "killing frosts" will decline. A killing frost is freezing temperatures that kill disease-carrying animals and insects. Those animals and insects will be able to move farther north (and south) in a warmer global climate into geographic regions where the diseases they carry are as yet unknown. Of continuing major concern is the mosquito. A warming climate will allow mosquitoes to move to higher latitudes. Mosquitoes carry such diseases as malaria, dengue fever, yellow fever, encephalitis, and the Nile River Virus.

Malaria is by far the most serious of these diseases. There are currently no reliable figures for how many people die each year from malaria. Estimates range from a low of 6 million to a high of 30 million people per year. Currently, 45% of the world's population resides within the zone of potential malarial transmission. If the Earth's climate warms as some predict, it is estimated there will be an additional 50 to 80 million cases of malaria worldwide.

The transmission of disease by mosquitoes is projected to rise by 60%

worldwide due to global warming. For example, the yellow fever mosquito was previously unable to survive at altitudes above 3,300 feet. This mosquito is now reported at a higher elevation in some parts of South America.

Global warming will result in a decline of air quality due to an increase in air pollutants, pollen, mold spores, and dust. This will increase the incidences of respiratory disease, asthma, and allergies. What happens to humans also happens to animals. A degradation of the environment from global warming will increase the number of animals carrying diseases infectious to humans. Rodents are of particular concern. Mice carry the Hantavirus, and rat fleas carry Bubonic Plague and typhus.

The deer tick carries Lime Disease, and if global warming increases the population of deer, Lime Disease could spread from its current geographic boundaries. All mammals can contract rabies, which thrives in warm climates. Longer summers could lead to an epidemic of rabies. Recent events in Europe demonstrate the susceptibility to disease of animals upon which we depend for food; these diseases include foot and mouth disease and Mad Cow disease.

Plants are no different. The potato blight in Ireland during the 19th Century caused mass starvation. A form of rust attacks wheat. The Med Fly destroys citrus crops.

The list is almost endless. If lack of adequate food reserves compounds the effects of global warming, causing a general reduction of health among human populations, the old scourges of mankind could return with a terrible virulence. Modern medicine made rare diseases like tuberculosis, polio, small pox, influenza, measles, diphtheria, leprosy, tape and ring worm, and pink eye…the list of the historical killers of mankind goes on.

Events of current history have not helped. Forced large-scale resettlement due to social unrest and war, increased drug resistance of bacteria due to indiscriminate use of antibacterial medication, higher human mobility through air travel, and the lack of insect control programs, all create opportunities for vector-born diseases. Disease finds further opportunity in various areas of the world where proper sewage disposal and clean water is not available. A disease to which there is no resistance or defense is the greatest threat of global warming to human beings.

As a result of global climate change, cholera, which is caused by human waste contamination in drinking water, is on the increase in Southeast Asia and several small Pacific Island nations. The big fear is that global warming will cause a mutation of a bacteria or virus to which the human population has no resistance. If that bacteria or virus is resistant to modern drugs, the mortality rate among the human population could be 95% or higher. Historical events provide evidence of this fact.

In 1776, when Captain Cook discovered the Hawaiian Islands, the native population of the Hawaiian Islands was estimated at 750,000 people. The native Hawaiians at the time of Captain Cook's discovery had no resistance to European diseases. Small pox, measles, cholera, and syphilis devastated the native population of Hawaii. That population was reduced from 750,000 people in 1776 to 22,000 people at the beginning of the 20th Century. That is a mortality rate of 98%. This was repeated among the North-American Indians as Europeans expanded their presence and their diseases in the Americas.

Another significant threat to human population is the inability to produce sufficient food. Either a severe drought condition or an overly abundant rainfall in the grain-producing regions of the world has the potential to cause worldwide famine. The combination of water and temperature changes poses significant problems for plants and animals. Forests, deserts, range lands, and other unmanaged ecosystems face stress if future global warming occurs. Their populations will decline, fragment, or become extinct.

On the North American continent, global warming will shift ideal temperature and precipitation ranges dramatically and rapidly to the north. This shift will be more rapid than that to which many species can adapt naturally. It will result in a decline in the biodiversity of many ecosystems.

The change in deserts will be more dramatic. As global warming changes climate patterns, there will be an irreversible process of desertification (drying soils and land degradation through erosion and compaction) in many regions. The activities of man will speed up the process. As grazing land becomes scarce, more pasture animals will be grazed on marginal land. As the animals graze on moisture-holding plants on the marginal land, the land will turn into desert. This is already happening at the edges of the Sahara Desert.

Global warming will cause a shift in the distribution of vegetation in mountains to higher elevations. Plant and animal life that exist only at high elevations will face extinction from the disappearance of habitat and the introduction of predators and diseases from lower elevations. Plant and animal life that live at high elevations in mountains are isolated from similar environments, and they are restricted in their ability to migrate.

Global warming will also affect coastal systems. Climate change, sea level rise and changes in storms and storm surges will erode the shores associated with coastal habitats. Other changes that will occur as the result of an incursion of seawater from storms and rising sea levels are: increased salinity of estuaries and freshwater aquifers; a change in the tidal ranges of rivers and bays; a change in the transport of nutrients for plants and animals; a change in chemical and biological pollution of coastal areas; and a change in flood patterns. The ecosystems at risk are water marshes, mangrove swamps, bayous, and other coastal wetlands, such as coral reefs, atolls, and river deltas.

Global warming will change the composition of the forests now found in temperate climate zones. Trees typically migrate 350 miles in 1,000 years. Many forest species may be forced to migrate between 100 and 350 miles toward the poles over the next 100 years. The projected temperature increase due to global warming over the next 100 years is on the average of 3.6° F. If the climate changes slowly enough, warmer temperatures may enable trees to colonize northward into areas that are currently too cold at about the same rate as southern areas become too hot and dry for them to survive.

This will, require that the trees to migrate at the rate of about 2 miles per year. Trees whose seeds are spread by birds may be able to migrate at that rate. But trees whose seeds are spread by the wind and nut-bearing trees such as oaks will not be able to migrate that quickly. (This does not take into account human activity that assists the movement of slow migrating trees.)

The range of a particular species of tree may tend to be limited to smaller areas as the southern regions of the United States warm. This will result in a loss of diversity. On the other hand, the projected increase in carbon dioxide from global warming may have a positive effect, since carbon dioxide is beneficial for fertilization and helps plants to use water more efficiently. This might enable some species of trees to resist the adverse effects of a warmer climate and drier soils.

A negative impact of increased warming and drier soils is the potential for forest fires. These will become more frequent and severe, as some areas of the United States have already experienced. Global warming will also change pest populations, increasing stress on trees.

The most complicating factor for the survival of any forest is that of whether global warming will make the region where the forest is situated wetter or drier. If the climate is wetter, the forest is likely to expand toward rangelands and other dry areas. If the climate becomes drier, then the forest will retreat from those areas.

Chapter 11
How High Is High

Changing climate due to global warming will also alter the geographic extent and plant composition of United States range lands. The grass rangelands comprise a large portion of the central part of the United States. This region does not currently receive enough rainfall to support forests, but does receive enough rainfall to support foliage for the grazing of cattle and sheep.

Availability of water in rangelands is the single most important factor in determining the value of the land for grazing. It is not known at this time if global warming will decrease or increase available water in the rangelands of the United States. A decline in available water will reduce the economic value of United States rangelands.

A wetter climate will enable forests to grow into areas that are now rangelands, while helping rangeland grasses to expand and grow into areas that are deserts. Elevated levels of carbon dioxide will introduce a shift from grasses to shrubs and other woody plants. Another possibility is that drier climate will cause a retreat of forests, and rangelands will take over where the forests once stood.

Some of the most biologically productive lands are Earth's non-tidal wetlands, which cover approximately 6% of the Earth's land surface. By definition, a wetland is an area that is flooded part of the time but not all the time. The high productivity of a wetland results from this characteristic. Flooding ensures that a wetland has ample supplies of water and nutrients. Wetlands are an important habitat for birds, fish and other species. They are also important mechanisms for cleansing pollutants from farming and other activities that runoff into rivers, lakes and streams.

Wetlands are broadly categorized coastal wetlands and inland wetlands. Coastal wetlands are mostly marshes and swamps that are flooded by ocean tides. Inland wetlands include wetlands along rivers and lakes as well as isolated wetlands not directly connected to any major body of water.

The impact of global warming on wetlands is uncertain, and it depends on changes in the amount of rainfall and when the rainfall occurs. Reduced rainfall and water tables will result in drier wetland areas for a longer part of the year.

A drier climate will also increase the use of irrigation by farmers. The use of local water for irrigation reduces groundwater tables. But if the water for irrigation is imported from elsewhere, groundwater tables and the amount of wetlands could increase. If precipitation increases, or if the severity of rainstorms increases, the resulting flooding may expand the area where flood plain wetlands form. If the runoff from increased rain also increases the rate at which silt is deposited, additional wetlands will be formed.

If the response to increased precipitation is to build dams, river levees, or other structures to prevent flooding, the total area where flood plain wetlands can form will decrease. The United States federal government has stepped into this debate by condemning and purchasing land in flood zones. This will reduce the building of flood prevention structures, and it will increase the area of wetlands—if there are no long periods of drought.

An issue related to inland flooding is that of coastal zone flooding due to rising sea levels. Both worldwide and United States coastal zones are threatened. The moderate scenarios for global warming predict a 2- to 4-foot rise in sea levels over the next century, and the less conservative scenarios predict an even higher rise in sea level over a much shorter period of time.

Coastal marshes, swamps, and bayous are especially vulnerable to rising sea levels because they are within a few feet of sea level. A rise in sea level will erode the outer boundary of these wetlands. New wetlands will form inland as those areas are flooded. However, the farther one moves inland, the steeper is the rise in elevation from the sea. As a result, the amount of newly created wetlands from rising sea levels will be smaller than the area lost. In the United States, a 2-foot rise in sea level will eliminate between 17% and 43% of all coastal wetlands. More than half of this loss will take place in Louisiana.

Many major coastal cities in the United States have low-lying areas. Boston, New York, Charleston, Miami, and Honolulu are examples. Other coastal cities in the United States actually have areas that are below sea level. Texas City, New Orleans, San Jose, and Long Beach, California, are examples.

The most threatened city in the United States by far is New Orleans. At 2 feet below sea level, it is an example of a city that sank after it was built. New Orleans protects itself against flooding with a series of bulkheads, dikes, levees, and parts of the delta that surround it. Although these structures do protect property, they cause beach erosion that keeps new wetlands from forming that will protect the city from a future rise in sea level.

In the United States, about 5,000 square miles of land is within 2 feet of high tide. A 2-foot rise in sea level will eliminate approximately 10,000 square miles of land in the United States, an area equal to the combined size of Massachusetts and Delaware. The vulnerability of coastal areas to severe storm damage will expand. The United States Federal Emergency Management Agency (FEMA) estimates that a rise in sea levels will increase the size of the once-in-100-year flood plane in the United States by 35% to 58%.

The purity of surface and ground water in coastal regions is threatened. The concern is increased water salinity. New York City, Philadelphia, and California's Central Valley obtain their fresh water from rivers that are only a few feet above sea level. If seawater is able to reach upriver due to rising sea levels, municipalities dependent upon these rivers for fresh water will find their intakes drawing in salt water.

Increased salinity will also affect aquifers situated near coastal areas. The Rartan-Magothy Aquifer in central New Jersey is one such aquifer. Shallow coastal aquifers found on small islands are also vulnerable. Freshwater is lighter than salt water. As a result, the freshwater that percolates down to these aquifers forms a "lens" of freshwater that floats on top of the salty groundwater. All freshwater in the State of Hawaii is obtained from such aquifers, as is the fresh water of many Pacific island nations. It is estimated that for every inch the sea level rises, a freshwater lens loses 40 inches of depth.

Global warming will affect the sea life of the coastal regions in other ways. It will exacerbate the anoxic conditions (too little oxygen in the water) of coastal areas. This will cause a northward migration of coastal fish. However, the ability of fish living in the Gulf regions of the United States to migrate further north is limited by the lower salinity in northern waters compared to Gulf waters. Some species will not be able to tolerate the change.

Inland freshwater fish will be adversely affected by global warming. However, man has already demonstrated in North America his ability and willingness to transport these fish to areas where they can survive. The impact of global warming on inland freshwater fish will be mitigated by the activities of humans.

Deep ocean fish have the ability to migrate to areas most favorable to their survival. Global warming will cause the fish to migrate northward and southward into cooler waters. Evidence suggests this is already occurring.

The Hawaii Monk Seal was in decline over the last 40 years. But marine biologists reported during the last 2 years that the Hawaiian Monk Seals in the northwestern Hawaiian Islands are better fed and in better health than previously observed. There can be only one reason for this. There is an increase in the population of fish in the latitudes north of Hawaii upon which the Hawaiian Monk Seal feeds.

The effect of global warming on the availability of food is a concern. Much of the world's food for human beings is derived from plants. In general terms, higher concentrations of carbon dioxide in the atmosphere and warmer climates will increase food crop yields, since carbon dioxide acts as a stimulant to plant growth.

Some recent experiments suggest that a doubling of carbon dioxide in the Earth's atmosphere will not increase the growth of plants over the long term. There is a problem here with resolution. No experiment on increased carbon dioxide in the atmosphere on plants has yet to be conducted over a period of more than 2 or 3 years.

An extended length of time is necessary to see what will happen to plants when they come into equilibrium with the increased carbon dioxide. Equilibrium is the change plants make to their biological structure to adjust to increased carbon dioxide. One possible change in this equilibrium is that plants will grow so fast in a carbon dioxide rich atmosphere that they will lose much of their nutritional food value.

There are long-term effects of global warming that might cancel out the benefits of carbon dioxide fertilization. Some of these are: an increase in plant diseases; an increase in the population of plant-eating animals (herbivores); and shading from an increased forest canopy (because the trees are also growing faster).

These changes in plant growth will force the world's human population to face new pressures. Total global food production is not expected to change substantially as a result of global climate change, and in fact, total global food production may decrease. Tropical and subtropical regions can expect a severe impact on food production from global warming due to reduced annual rainfall and changing climate patterns. The flexibility in crop distribution and the variety of crops grown in these regions is expected to decline.

Increased atmospheric carbon dioxide, higher temperatures, and increased precipitation are not expected to impact overall crop production in the United States. A study of wheat production in Kansas (and elsewhere) indicated that rising carbon dioxide levels would be beneficial to that crop. A higher concentration of atmospheric carbon dioxide will allow wheat to grow larger and healthier, produce a higher yield, use water more efficiently, and be tolerant to drought. This is important since global warming will bring on long, hot summers that will dry the soil.

Dry soil is of particular concern because of its effect on crop yields, groundwater resources, lakes, and river ecosystems. Temperature, solar radiation, atmospheric humidity, and the wind all affect how dry the soil will be. The combination of these four factors causes the soil found between the north and south latitudes of 20° and 32° to be extremely dry. An increase in global temperature of 6.5° F will quicken the evaporation from the soil in these regions by 30% to 40% while annual precipitation will increase only by 10% to 15%. The result is an expansion of land areas deficient in precipitation toward both the poles and the equator.

Drought is a concern for all farmers, and it is a particular concern for farmers of the Midwestern and southwestern regions of the United States. Global warming will change the precipitation patterns in these regions that over the last several thousand years have suffered a number of long and severe droughts.

During the 20th Century, the United States suffered two major droughts, the classic Dust Bowl of the 1930s and another in the 1950s. During the 1950s, the Great Plains and the southwestern United States withstood a 5-year drought, and in 3 of those years, drought conditions stretched coast-to-coast across the United States. Texas rainfall decreased 40% between 1949 and 1951. By 1953, rainfall in Texas was 75% below normal. Excessive temperatures heated cities like Dallas, where the temperature exceeded 100° F on 52 days during the summer of 1953 (a record that was broke in 2000). Similar effects were felt all across the Midwestern United States.

No part of North America is immune from drought, and drought conditions occur frequently all across the North America. A 5-year drought in the 1960s drained the reservoirs in New York City to 25% capacity. In 1988, Florida, Oklahoma and Texas suffered drought and forest fires that caused $500 million in damages. (At its peak, this drought covered 36% of the United States, and it was at the time the most expensive natural disaster of any kind to affect the U.S.) That same year, Canada suffered its fifth highest forest fire season in 25 years.

There was a drought in the Midwest in 1997. The West Coast of the United States experienced a 6-year drought from the late 1980s to the early 1990s. In 1998, Oklahoma suffered a devastating drought, and starting in that same year, 3 years of record low rainfall plagued northern Mexico. The worst drought for northern Mexico in 70 years was declared in 1998. In 1999, spring rainfall fell to 93% below normal. The government of Mexico declared 5 northern states disaster zones and increased that number to 9 in 2000.

During the same period of time, Florida, Texas and the northwestern region of the United States were also struck with severe drought. Forest fires in the northwestern United States during the summer of 2000 were declared out of control for over two months. Fire fighters were brought in from as far away as New Zealand and Australia.

As devastating the droughts were for the 20th Century North American continent, they pale when compared to the geological records of droughts in the distant past. North America has suffered reoccurring droughts over the last 8,000 years. Evidence suggests that some lasted between 50 to 400 years. Such lengthy droughts are thought to have contributed to the destruction of both the Mayan civilizations of Central America and the Anatazi Indians in the southwestern United States.

The business of agriculture in the United States has learned to adapt to these extensive climate changes. Today's agricultural communities in the United States are among the most advanced in their use of technology. Further, United States' agriculture is regionally diverse, highly productive, and intensively managed. While specific regions may suffer adversities due to global warming, the impact of global warming is not expected to significantly influence this industry as a whole. It must be emphasized again that what is described here are projected impacts that will occur in conjunction with global warming. There is nothing at this time to suggest that these impacts have occurred. They are, rather, predictors that, if observed, may suggest global warming has arrived.

It is also emphasized that effects of global warming may not be those predicted. Nature has the all-too-common habit of producing unexpected results. This fact has created the intense debate over what is exactly happening regarding global climate change.

Chapter 12
It's Not So Much What You Say

The Iroquois League faced a decision in the year 1776. This confederacy of six Indian tribes—the Mohawk, Oneida, Onondaga, Cayuga, Seneca, and Tuscarora—claimed the Ohio Valley land directly west of the thirteen fledgling British colonies on the East Coast of North America. Those thirteen colonies were at war with their mother country, Great Britain. The question the leaders of the Iroquois had to answer was which side to join? The Iroquois leaders were well aware of the strength of the Europeans, and they knew the cost of a bad decision.

The Iroquois League was second to none among the Indian tribes of North America in political organization, statesmanship, and military ability. During the 17th Century, the Iroquois League, equipped with firearms provided by the Dutch, embarked on a war that eliminated the Huron, the Tobacco and the Neutral Nations, the Eries, the Conestogas, and the Illinois Indian Tribes.

By the end of the 17th Century, the Iroquois League reached the height of its power. Its population was approximately 16,000 people, and these people controlled the territory bounded by the Kennebec River, the Ottawa River, the Illinois River, and the Tennessee River. However, the conquests of the Iroquois League were checked in the west by the Ojibwa, in the south by the Cherokee and the Catawba, and in the north by the French settlements.

The Iroquois League was the balance of power between the French and the British during the 1765 French and Indian Wars. The Iroquois League allied itself with the British. The one exception was a portion of the Mohawk Indian Tribe who converted to Christianity through the efforts of French Jesuit missionaries. Those Mohawks, called French Mohawks, fought for the French against the British. This division was the beginning of the end for the Iroquois League.

When the British defeated the French, the attitudes of the British changed toward the Iroquois League. The British decided not to treat the Iroquois League as a sovereign nation. This led to a disastrous decision by the Iroquois League during the American Revolution. The political structure of the Iroquois League was such that any decision for war required a unanimous decision of all the tribes in the League. Because of the British decision not to treat the Iroquois League as a sovereign nation, the Iroquois League decided to remain neutral during the America Revolution.

Each tribe decided for itself which side it wished to support. All the tribes of the Iroquois League, with the exception of the Oneida Tribe, decided to support the British. They participated in raids on New York and Pennsylvania along side the British Rangers of Water Butler that resulted in the Cherry and Wyoming Valleys Massacres.

The American colonies won their war of independence against Great Britain when the Treaty of Paris was signed in 1783. The members of the Iroquois League who allied themselves with Britain did not recognize the treaty, and they continued to fight. The newly formed United States took action, defeated the League, and, in the Second Treaty of Fort Stanwix, officially disbanded the Iroquois League in 1784.

The one exception was the Seneca Indian Tribe, the tribe that occupied the westernmost lands of the League. The Senecas were able to hold out against an onslaught of military action, destruction of crops, and smallpox-laden blankets until they too capitulated in 1789. From this point on, the tribes of the Iroquois League descended into a disease-ridden population aggravated by severe alcoholism. By 1799, the population of the formerly powerful Iroquois League was less than 4,000 people.

What happened to the Iroquois League should be a lesson to us today when we consider the question of global climate change. The worst thing we can do is nothing. The second worst thing we can do is make a decision based on faulty information. There are four arguments regarding global climate change. They are: It is happening; it's not happening; it's happening, but the cause is a source other than greenhouse gas emissions; and, it's happening, but the impact of the warming will cause results other that those predicted (the position here).

What is demonstrated by the arguments for and against global climate change is the current uncertainty regarding a number of issues. This uncertainty is represented by the questions asked about global climate change, such as: How much warming has occurred due to the increases in atmospheric gases introduced by humans? How much warming will occur in the future? How fast will the warming occur? What other kinds of climate change will be associated with future global warming? Here we look at the reasons why some say global climate change is not occurring.

Predictions of Rapid Warming Are Wrong

The Earth's climate is complex and chaotic. There is no way to know what might have been, had humans not introduced greenhouse gases into the Earth's atmosphere. What is occurring with the Earth's climate may be due to human influence, but it is more likely that the change is simply another natural cycle.

The global warming observed during the 20th Century is no more than a normal climatic variation. Ninety-eight percent of total greenhouse gas emissions originate from natural sources such as water vapor. Manmade sources account for only 2% of all global greenhouse emissions.

Manmade emissions have only a small impact on the global climate. Though global climate has warmed over the last 100 years, 70% of that warming occurred before 1940, but nearly all of the increase in greenhouse gas emissions occurred after 1949.

Further, as much as 30% of the warming that was observed in the 20th Century may be a rebound from the preceding 400-year period of cooler global temperatures known as the Little Ice Age. There is clearly more going on with the Earth's global climate than a simple, direct response to greenhouse gas emissions.

The United Nations Intergovernmental Panel on Climate Change (IPCC) is the principal advocate of global warming. In 1990, the IPCC predicted an increase in global warming of 6° F by the year 2000. Three years later, in 1993, the IPCC adjusted its prediction downward to 2° F by the year 2000. This 2° F increase in global temperature is well within the historical 3° F to 4° F range of the Earth's normal climate temperature variation. The IPCC has admitted its predictions in 1990 were wrong.

Ground-based instrumental measurements of the Earth's global temperature indicate that the Earth warmed by about 1.5° F between 1860 and 2000. This record is not long enough (lack of resolution) to determine if this warming is from natural variation or is due to human activities. Observed changes in global warming during the last decade of the 20th Century are several times lower than those projected by climate models.

Observed warming was confined mostly to Siberia and the northwestern part of North America. Historical geophysical data does not support a claim for a catastrophically warmer global climate. Studies of ice cores, tree rings, and ocean sediment suggest there were many periods in the Earth's recent past when global climate was considerably warmer than it is today without detrimental effects. It is inaccurate to imply a statistically significant trend in global climate change based on a few decades of temperature measurements. There were two general cycles of warming and cooling recorded in the United States over the last 100 years. Those cycles show no major trend towards global warming.

Carbon dioxide in the Earth's atmosphere has increased for the last 18,000 years. Humans played no significant role in that increase. Today, the total annual contribution of humans to atmospheric greenhouse gases lies somewhere between 0.1% and 0.2%.

Of the 186 billion tons of carbon dioxide that enter the Earth's atmosphere each year, only 6 billion tons are from human activity. Approximately 90 billion tons come from the biological activity of the Earth's oceans and another 90 billion tons come from such sources as volcanoes and decaying land plants.

At its present level, carbon dioxide is a minor constituent of Earth's atmosphere. It comprises less than 4% of all gases present. According to the historical record, the Earth is currently carbon dioxide impoverished.

Chapter 13
Data Error

Global climate cycles of warming and cooling are natural events spanning many hundreds of thousands of years. It is doubtful that these cycles will stop because of greenhouse gas emissions by humans. Errors in data collection and interpretation are what led to erroneous conclusions about global warming. The following are some examples how the errors in data collection and interpretation occurred.

Heat Island Effect

Urban areas are warming more rapidly than rural areas. These two areas use land differently. Rural areas are covered mostly with vegetation, while urban areas are covered mostly with buildings, concrete, and asphalt. Energy consumption in rural areas is lower than in urban areas. As a result of these differences, urban areas generate more heat. The increased heat generation that takes place in densely developed areas is known as the "urban heat island" effect.

Cities and other urban areas have tall buildings with vertical walls. The vertical walls act like light traps, capturing and storing heat during the day, and re-emitting that heat at night. This is what causes urban areas to be warmer than the surrounding country. Airports also generate an urban heat island effect. Airports have large, paved runways that soak up heat from the Sun. Enormous amounts of fuel are combusted at airports. Both the runways and burning fuel keep airports warmer than the surrounding country.

Both cities and airports artificially increase atmospheric temperature in their immediate locality. This is important to know, because many surface stations measuring atmospheric temperature are located in either cities or at airports. Those stations are reporting an increase of global temperature not due to increased greenhouse gases, but due to increased urbanization and expansion.

Even if a temperature recording station is located in a rural area, it can still record erroneously high temperatures. A rural station situated in a field surrounded by a line of trees 30 or more feet high and less than 150 feet away, the temperature at the station will be 0.5° F higher than normal. The trees are too close to the measuring station to allow for the free and clear flow of air needed for an accurate temperature reading.

Rising Sea Levels

It is argued that global warming will result in rising sea levels. A measurement of rising sea levels requires a determination of a concept known as global "mean sea level." That is, what is the average level of the Earth's oceans? A serious problem confronts any person who seeks to determine the Earth's mean sea level. This determination requires a comparison to a fixed object.

One of the oldest gauge record in the world, often cited as proof of a rising sea level is located in Aberdeen, Scotland. The Aberdeen Tidal Gauge was established in 1862. There are problems associated with this tidal gauge. It is on a pier, and over the years such a structure has a tendency to sink, giving a false appearance of a rise in global mean sea level.

Also, the mantel below the Earth's surface began to force the Earth's crust upward when the dead weight of the ice from the last Ice Age was removed. This too will provide a false reading of global mean sea level. Because of these two variables, the conclusion is that the Aberdeen Tidal Gauge is not an accurate measurement of global mean sea level. This may change in the future when it possible to calibrate these influences on the tidal gauge. Until then, the Aberdeen Tidal Gauge cannot be considered a reliable indicator of change in sea level.

Another theory suggests warming oceans will result in more evaporation and thus more cloudiness. The increase in clouds will result in an increase in precipitation over the Polar Regions, which will build up the polar ice caps. That will cause a decrease in the global mean sea level from global warming as more and more water is stored in polar ice, and the predicted rise in sea level will not occur.

There is evidence indicating this is exactly what is happening. On July 1, 1841, the Ross-Lamprier benchmark was inscribed by Capt. Sir James Clark Ross on a cliff near Point Puer on the Isle of the Dead in southeastern Tasmania, Australia. The purpose of the inscription was to mark a "zero" point of the sea, or the mean sea level. In the 160 years since the inscription of the mark by Capt. Ross, that mark is now nearly 20 inches above the water at mid-tide. The Isle of the Dead did not have large glaciers of the last Ice Age. Geological rising of land cannot account for the change. Thus, this benchmark disclaims a rise in mean sea level during the 20th Century.

Satellite Measurements

NASA's TIROS weather satellites provide the most accurate measurement of the Earth's global temperature to date, covering 97% of the Earth. These satellites record that the Earth's global temperature cooled slightly since 1979, and that is a contradiction to the prediction of global warming. These satellites are not subject to the urban "heat island" effect that influences the temperature measurements of ground-based thermometers situated in major cities and airports.

Current calculations by satellite measurements indicate that the Earth is warming at the rate of 0.5° F per decade, an amount that is considerably less than the rate of warming measured by ground-based thermometers. The satellite measurements are corroborated by atmospheric balloon measurements.

Satellite data on global temperature extend back to 1979, and weather balloon data on global temperature extends back to 1960. Neither method of measurement shows a trend towards global warming. Surface temperature measurements found 1997 to be the warmest year on record. Satellite and weather balloon measurements rank 1997 as the 7th coolest year since satellite measurements began.

Satellite measurements cover a larger portion of the surface of the Earth. Thirty percent of the Earth's surface is land, and as such, land-based stations measure temperature over less than 1/3 of the Earth's surface. Deserts, mountains, forests, and oceans form vast areas that have few or no stations recording temperature. By contrast, satellites cover 99% of the Earth's surface.

Further, measurements of global warming taken at the Earth's surface are not really global. There are approximately 2,900 surface weather stations worldwide measuring atmospheric temperatures. Only 161 (5.5%) have continuously recorded atmospheric temperatures since 1900, and all but 19 of these 161 stations are located in the United States. This is a problem for the detection of a long-term trend. Changes in climate temperature are given greater weight within the United States than change in climate temperature occurring elsewhere on Earth.

It is estimated that at least 8 billion temperature sensors are required to get measurements of the Earth's global temperature to an accuracy of 1°

F. This estimate is based on the approximation that the surface of the Earth is 200 million square miles. There are currently approximately 7,000 surface-based stations measuring the Earth's atmospheric temperature, and in some places that temperature can vary 50° F in one day and 100° F in one year. This does not even begin to measure the 70% of the Earth that is covered by water.

Limited Global Variation

Not all variations in climate are global. Only about 8% of climate variations can be categorized as global. The remaining 92% are comprised of regional variations. If the global variations are removed from the equation, there will still be significant local and regional variations occurring in the climate or, more properly stated, in the weather.

Resolution

There is also the problem of resolution. Depending upon the base year, or the interval chosen, a different history of climate change will result. This different coverage leads to a different construction of global climate. The proponents of global warming chose the 20th Century for the interval, and this enhanced the data for global warming. A longer timeline demonstrates that the Earth's global temperature today is nowhere close to historical highs.

Computer Model Projections

A computer model for weather prediction is simply a computer program designed to project future weather based on currently known information. Computers modeling future climate have two inherent weaknesses. The first is that current computer models do not include all the various factors influencing global climate change. The second is that the accuracy and reliability of the data used for the prediction of future global climate change is not known. This is a classic case in computer programming of "garbage in equal's garbage out." The results from computer modeling are only as good as the data used to make the computations. To date, the data used are not reliable.

Computer models predict that global warming will occur during the 21st Century. However, those models are limited. The predictions are based on General Circulation Models (GCMs), complex computer programs that attempt to simulate the Earth's atmosphere. They do not take into full consideration the effects of clouds, precipitation, the oceans, or the Sun. The influence of each on global climate changes is still not well known, though each plays a major role.

One of the reasons global temperature variations are difficult to model is that they are sensitive to unusual circulation patterns, which occasionally cause them to follow a trend opposite of global warming. This is where the major deficiencies of GCMs occur. For example, clouds have a significant influence on the Earth's atmospheric heat budget. But the physical processes that form clouds and determine their characteristics are on a scale too small to be accounted for directly by GCMs.

Further, the factors that cause a major influence on global climate—temperature, precipitation and storms—are so interrelated that it is impossible to understand one independent of the others. For example, the cycle of evaporation and precipitation transfers not only water from one place to another, it also transfers heat. Computer GCMs do not currently address such complexities.

Again, GCMs are only as good as the data entered into their programs. For example, many GCMs assume clouds absorb 3% of the Sun's radiation, and there is an increase in greenhouse gases of 1% per year. Recent studies indicate that clouds absorb 19% of the Sun's radiation, and historical data show an increase of greenhouse gases of only 0.3% to 0.4% per year. These two assumptions in the design of GCMs caused an error of 60% in the projected increase of greenhouse gases and an error of more than 600% in the projected rate of solar radiation absorption by clouds.

Uncertainty about the ability of GCMs to simulate natural climate variability remains a problem. GCMs omit many variables that could result in long-term changes in the Earth's global climate. GCMs do not, for example, include the effects of volcanic eruptions, which can cool the global climate temporarily by several tenths of a degree. In order for GCMs to accurately predict future global climate changes they must include provisions for the behavior of the oceans, vegetation, the cryosphere (sea ice, glaciers, and ice caps), and the output of the Sun. This is something GCMs currently do not do.

El Niño

Global warming does not cause the El Niño Southern Oscillation (ENSO, referred to here as "El Niño"). Up to now, there is no apparent connection between the two. There is clear evidence from a variety of sources that El Niños have existed for hundreds, and maybe thousands, of years. There are also clear indicators that warmer ocean surface temperatures will enhance the effects of an El Niño event. The global warming pattern since 1970 look more like the footprint of an El Niño event than like global warming.

It was recently determined that the climate patterns of the Pacific Ocean, in addition to the El Niño events, also go through a longer 40-year oscillation of change, called the Pacific Decadal Oscillation. This either enhances or decreases the effects of an El Niño event. It is very possible that the current observation of global warming is simply a naturally occurring oscillation of Pacific Ocean climate patterns that are increasing the influence of El Niño.

Personal Agendas

There are personal agendas that throw suspicion on the claims of global warming. There is a lot of grant money available for research on global warming. A researcher funded by government or corporate grants has an incentive to look for evidence that supports the position of the sponsor in order to maintain funding for research.

Support for the idea of global warming has the potential to generate additional tax revenues for governments. The tax will be on energy consumption. It is potentially a large source of additional revenue, and it can only be supported if the public is convinced global warming is happening.

A considerable number (but not all) of the advocates for global warming fall into one of these two categories. This is unfortunate because it masks what is really happening with global climate change. There are other possibilities, and many scientists are doing good work on those possibilities. What is puzzling some of those scientists the most is why global warming is not coming about as predicted.

Chapter 14
Marine Worms and a Question

H.G. Wells wrote in War of the Worlds that the most insignificant among us have saved us, referring to the microbes that killed off the Martians. Sometimes the insignificant is significant. How true that is when a consideration of global climate change drops, literally, to the depth of marine worms.

Marine worms, more properly referred to as polychaetes, populate the depths of the Earth's ocean floors. They are an ancient marine animal found throughout the oceans of the world from the sand at the edge of the surf to great ocean depths. Marine worms inhabit almost all available space on the ocean bottom, and they range in length from a tenth of an inch to over 200 feet. It is estimated that there are 45,000 species of polychaetes, some of which can be traced back to the Cambrian Explosion 500 million years ago.

Polychaetes play an important role in Earth's global climate. They convert organic debris that falls to the ocean floor into carbon dioxide. The carbon dioxide is dissolved in the ocean's water, and that water then rises to the ocean's surface. Marine plankton use the carbon dioxide generated by the polychaetes to produce food and energy for themselves through photosynthesis, in the process, releasing oxygen into the Earth's atmosphere. It is thought by some that as much as 1/3 of the Earth's oxygen is generated through this process.

It is this connection between the generation of carbon dioxide by polychaetes and oxygen producing plankton that makes some segments of the scientific community nervous when it comes to the current hype about global warming. The argument for global warming is based on an assumed correlation. Yet the predicted global warming has not arrived. Some concerned scientists are well aware that another mechanism might be at work, like the marine worms, that will produce an unexpected result. The scientific community has focused on global warming for almost 30 years.

However, a small group of scientists faced with the facts presented by those who say global warming is not occurring, and understanding the possibility of other influences asked the question, "Why is global warming not occurring as predicted?"

Those scientists took a look at the science that predicted global warming, and they found the science good. That is, the scientists who made the predictions followed proper scientific methods in making their predictions.

Science can be a powerful predictive tool. Much can be learned when science makes a correct prediction, and just as much, and sometimes more, can be learned when science makes a mistake. This is what caused some scientists to investigate the failure of the accuracy in predicting global warming.

Those scientists concluded that something is influencing the Earth's albedo. Albedo is the amount of light reflected by a geographic area, such as land, water, and ice. It is a measurement of the ratio of reflected light to incoming light. Albedo can be expressed as either a percentage or a fraction of 1, and the higher the number, the greater is the amount of light reflected. The aggregate albedo for the Earth is approximately 0.3. That is, the Earth on the whole reflects approximately 30% of all light that strikes it.

A change of just 1% in the amount of light the Earth reflects, or absorbs, could cause either global cooling or global warming. Ice reflects back approximately 90% of the light that strikes it. This is one of the factors that keep areas covered by ice cool. Sea ice can appear and disappear quickly under the right circumstance, and it can influence a rapid change in climate.

On the other hand, ice sheets like those in the Antarctica and Greenland, because of their sheer size, take longer to develop and melt. It is possible that ice sheets could help bring about a change in global climate through a rapid surge outwards into the ocean, creating a large number of icebergs that reflect back the Sun's heat and cool the global climate.

Current observations reveal that mountain glaciers are in recession, ice sheets are melting, and sea ice is in retreat. This should contribute to global warming because darker seawater and soil does not reflect back as much light as ice. A side effect of this is less sea ice and fewer icebergs, resulting in more open ocean water. The increase in exposed ocean water may allow the oceans to absorb more carbon dioxide, and this might be the reason global warming is not occurring as predicted.

In close conjunction with melting ice is the Earth's soil. As the Earth's land surface frees itself from snow and ice due to global warming, more of the Earth's soil is available for carbon sequestration, the holding of carbon dioxide by soil and plants. On a global scale, soils are a more important reservoir of carbon than vegetation.

The major store of carbon in the Earth's soil takes the form of heterogeneous organic compounds derived from decaying plant material. This supports a biomass of bacteria and fungi that is an important source of carbon sequestration. It is suggested (but not yet proven) that the current levels of carbon dioxide in the Earth's atmosphere caused an increase of 30% to 40% in the amount of carbon stored in the Earth's soils. This might be the reason global warming is not occurring as predicted.

Dust also influences the Earth's albedo. The presence of dust in the Earth's atmosphere prevents light from the Sun reaching the Earth's surface. The Earth's deserts are expanding, and mankind is clearing more land for farming. Mineral dust is a natural product. However, the rates of dust emissions have increased recently, due to drought compounded by mankind's land use practices, for each year over the last 20 years.

Satellite measurements found large clouds of dust scattered across northern Africa and parts of Australia. Dust from the Earth's great deserts can migrate on winds around the world. In 2001, Seattle received dust from the Gobi Desert, and in the same year, Florida received dust from the Sahara Desert. This dust is a dominant factor in scattering light over the Earth's oceans.

This causes the atmosphere to cool. The cooling, in turn, may inhibit plant growth, which creates more land susceptible to erosion, creating more dust. Dust particles reflecting the Sun's light can counteract the effects of greenhouse gas emissions. This might be the reason global warming is not occurring as predicted.

The cooling influence of aerosols was previously discussed in brief. The role aerosols play in the cooling of the Earth's atmosphere is a major source of uncertainty. Within the past several years some scientists began to look at the implications aerosols have on global climate, and they concluded that aerosols are the second most influential factor on global climate change caused by humans.

The greenhouse effect is caused by the emission into the Earth's atmosphere of carbon dioxide and other gases. Industrial nations also produce sulfur that is emitted into the Earth's atmosphere. Sources of sulfur emissions are coal-fueled electrical generation plants, automobiles, slash-and-burn agricultural practices, and forest fires.

Sulfur is the source of acid rain. The sulfur mixes with water in the atmosphere, creating sulfuric acid. It is from this creation of sulfuric acid that the name "acid rain" is derived. However, before sulfur emissions become acid rain, the sulfur is first converted into very small particles of sulfate. These particles are called aerosols.

Aerosols influence the Earth's albedo. When manmade and natural aerosols are released into the Earth's atmosphere, the aerosol particles reflect the Sun's radiation back into space like trillions of microscopic mirrors. This cools the area of the Earth's surface over which sulfate aerosols are generated.

An important effect of aerosols on the Earth's atmosphere is what happens when the aerosols are mixed with clouds. Certain aerosols such as sulfate and sulfuric acid react with atmospheric water vapor to become what is called "cloud condensation nuclei." Large numbers of condensation nuclei in clouds produce many small water droplets. Clouds composed of such small droplets are highly reflective to incoming solar radiation, and they reflect more of the Sun's radiation back into space than clouds without the condensation nuclei. Such clouds also persist longer, causing further cooling.

The reflective qualities of aerosols appear on the surface to be a good explanation of why global warming has not occurred as predicted. As is so often the case when discussing global climate change, there is a problem with this explanation. Unlike greenhouse gases, which persist in the Earth's atmosphere over many years, aerosols persist in the Earth's atmosphere less than a week.

Consequently, aerosols are concentrated near sulfur-producing, industrial areas. Manmade sulfur emissions that cause aerosols are emitted into the Earth's atmosphere on an order of 2 to 3 times the rate of natural sources of sulfur emissions. However, the short persistency (lifetime) of aerosols in the Earth's atmosphere does not permit them the opportunity to mix with the Earth's global atmosphere on the whole.

Enter again our old friends from the Snowball Earth Theory, the volcanoes. Volcanoes have both a short-and a long-term effect on the Earth's climate. For example, the Earth's average global temperature dropped about 1° F in 1991 for approximately 6 months after the eruption of Mount Pinatubo in the Philippines. In Indonesia, the eruption of Tambora in 1815 caused very cold temperatures across North America, resulting in two years of crop failures and famine. Tree ring evidence suggests that the cold summers of 1601, 1641, 1783, 1816, and 1912 were associated with known volcanic eruptions.

Volcanoes affect the Earth's climate through the gases and dust particles thrown into the atmosphere during eruptions. Volcanic dust and gases cool or warm the Earth's land and oceans, depending upon the interaction of the Sun's radiation with the volcanic material. Volcanologists believe that the Earth's ability to maintain a relatively mild climate for millions of years is due to ongoing volcanism.

Volcanic dust injected into the Earth's atmosphere causes a temporary cooling. The amount of cooling depends on the amount of dust put into the Earth's atmosphere, and the duration of the cooling depends on the size of the dust particles. Smaller dust particles tend to stay in the Earth's atmosphere longer. Dust particles thrown up into the lower atmosphere are quickly washed out of the air by rain. However, dust particles injected by an eruption into the dry Stratosphere can remain there for weeks to months before they finally settle. These particles block the Sun's radiation and cause cooling over large areas of the Earth.

The effects of volcanic dust can be far-reaching. When Mt. St. Helens erupted in 1980, the ash cloud rose to over 50,000 feet, and its ash deposits were inches deep in cities as far away as Washington, D.C. The Mt. Pinatubo eruption created a volcanic cloud of ash that covered 42% of the Earth's surface within 2 months.

When Tambora on the Island of Sumbawa, Indonesia, erupted in 1815, a 13,000-foot mountain with a diameter at sea level of 38 miles was reduced to a crater 3,640 feet deep. It was the largest volcanic eruption in recorded history.

The Tambora eruption emitted 31.5 cubic miles of ash in a column approximately 26.5 miles (140,000 feet) high. Global winds swept the fine dust particles around the Earth within a matter of months, resulting in devastating climate changes. An article entitled "Melancholy Weather" from the North Star Newspaper of Danville, Vermont, dated June 15, 1816, provides insight into how a volcano in Indonesia can influence weather in the northeastern United States:

> Some account was given of the unparalleled severity of the weather. It continued without any

essential amelioration, from the 6th to the 10th instant-freezing hard five nights in succession as it usually does in December. On the night of the 6th, water froze an inch thick-and on the night of the 7th and the morning of the 8th, a kind of sleet or exceeding cold snow fell, attended with high wind, which measured in places where it was drifted, 18 to 20 inches in depth. Saturday morning, the weather was more severe than it generally is during the storms of winter.

It is estimated that 92,000 people were killed by the Tambora eruption. Ten thousand people died from bomb impacts, tephra fall and pyroclastic flows. An additional 82,000 people died by starvation and disease that resulted from the change in global climate caused by this volcanic eruption. In the past 200 years, there were 6 such major eruptions, but of the 6, only Tambora had a negative effect on global climate.

Volcanoes that eject large quantities of sulfur compounds affect the Earth's climate significantly more than those volcanoes that eject just dust. The resulting sulfuric acid can cause significant cooling of the Earth's surface for up to two years after an eruption. Sulfuric acid in the stratosphere is believed to be the primary cause for the global cooling that occurred after the eruption of Tambora and Pinatubo.

Volcanic eruptions also release large amounts of water vapor and carbon dioxide into the Earth's atmosphere. Both water vapor and carbon dioxide are powerful greenhouse gases. It would appear, therefore, that a major volcanic eruption holds the potential to cause temporary global warming instead of global cooling. This has not proven to be the case.

The Earth's atmosphere already contains a large quantity of carbon dioxide and water vapor, and a major volcanic eruption does not change that amount by very much. The water vapor condenses out of the Earth's atmosphere as rain within a few days, and the carbon dioxide quickly dissolves in the ocean or is absorbed by plants. Current thought about water vapor and carbon dioxide from volcanic eruptions is that they do not have a long-lasting influence on global climate because right now there are too few major eruptions. However, as the natural biological sinks for the absorption of carbon dioxide are consumed, or if volcanic activity increases, this could change in the future.

That said, volcanoes are still very much with us. The global effects of the sulfur ejected from a volcano have as much to do with the latitude at which the volcano exists as with the magnitude of the eruption. There seems to be a direct correlation between the latitude of the volcano and the impact the emissions of that volcano has on global cooling. This relationship is based on the fact that the elevation between the volcano's summit and the distance to the Stratosphere decreases with an increase in latitude.

The height of the stratosphere above the Earth's surface decreases the closer one gets to either of the Earth's Polar Regions. The summit of a volcano in an extreme north or south latitude is closer to the Stratosphere than the summit of a volcano located on the Earth's equator. As such, a smaller volcanic eruption at higher latitudes can emit just as many sulfur compounds into the Stratosphere as a larger volcanic eruption near the Earth's equator.

The conclusion of the scientists asking the question, "Why?", is that both manmade and volcanic generated aerosols are masking global warming. Under certain conditions, aerosols reflect back into space more than twice the energy generated by greenhouse gases. It appears that aerosols in the Earth's atmosphere have the potential to cool the global climate by as much as 0.7° F to 1.5° F provided nothing else changes in the future. It is the increase in the Earth's albedo from the existence of aerosols in the atmosphere that is the reason global warming is not occurring as predicted.

Chapter 15
Star Light, Star Bright

One of the intriguing things about the study of global climate is that out of obscurity can be derived an observation of events so obvious one wonders why it was not noted earlier. Milutin Milankovitch (1879-1954) made such a discovery. Outside of the field of astrophysics his name is almost unknown. Yet, Milankovitch developed a significant theory relating the motion of the Earth through space and long-term climate change.

Milankovitch was born in the rural village of Dalj, Serbia. He attended the Vienna Institute of Technology where he graduated with a doctorate degree in Technical Sciences. He worked for a short period of time as the chief engineer of a construction company. Then, he accepted a faculty position in applied mathematics at the University of Belgrade. It was a position he held for the remainder of his life.

Milankovitch dedicated his career to developing a mathematical theory of climate. He theorized that the ice ages were related to changes in the Earth's orbit. During the 1920s and 1930s, Milankovitch hypothesized that the gravitational forces of the Sun and other planets of the Solar System subtly influence the Earth's rotation around the Sun and the Earth's tilt. These subtle changes result in different distributions and intensities of sunlight striking the Earth, causing either a warming or a cooling of the Earth's global climate over tens of thousands of years.

Milankovitch found that the Earth's orbit is elliptical, and it varies between an ellipse and a circle over a 95,000-year cycle. He also found that the Earth's axis wobbles on a 26,000-year cycle, and that the tilt of the Earth changes every 41,000 years. Taken together, these periods of the Earth's orbital motions are known as Milankovitch Cycles. Milankovitch Cycles account for some of the past changes in the Earth's global climate.

It is difficult for a person not trained in the perception of three dimensional orbital mechanics to visualize the different orbital motions of the Earth through space. First the Earth's orbit around the Sun changes from an ellipse (oval) to a circle and back again over a 95,000-year cycle. This is called the Earth's eccentricity. When the Earth's orbit around the Sun is most elliptical, there is as much as a 30% difference in the distance between the time it is the closest to the Sun (perihelion) and the time it is the farthest from the Sun (aphelion).

Today, the Earth's orbit around the Sun is almost circular. Perihelion occurs on January 3 when the Earth is 91.5 million miles away from the Sun, and aphelion occurs on July 4, when the Earth is 94.5 million miles away from the Sun. A quick mind will note that for the Northern Hemisphere, the Earth is farther from the Sun in the summer than it is in the winter. Therefore, climate should be colder in the summer and warmer in the winter based on the Earth's eccentricity.

However, it is not, and the reason it is not is due to the Earth's tilt. The tilt of the Earth's axis is called obliquity. Visualize the Earth's axis as a pole standing upright at 90°. Bend the pole over from 90° to an angle of 67.55°. This is the obliquity, or angle at which the Earth is tilted in relation to the Sun. At this angle, the sunlight striking the Earth comes in at a greater angle during some parts of the year than others. The greater the angle, the less is the radiation from the Sun that strikes the Earth's surface. It just happens that the Earth's tilt is such that the angle is greater during the winter months and less during the summer months for the Northern Hemisphere.

As a result, even though the Earth is 3 million miles closer to the Sun during the winter months, the greater angle at which the Sun's radiation strikes the Earth's surface results in colder weather for the Northern Hemisphere during winter months. Less of an angle than the Earth's current 23.45° tilt means fewer seasonal differences between the Earth's Northern and Southern Hemispheres, and a greater angle results in greater seasonal difference between the two hemispheres.

The Earth wobbles as it spins, just like a top does when its spin begins to slow down. As a result of this wobble, the Earth's tilt in relation to its orbit around the Sun changes on a 12,000-year cycle. This is called precession, or precession of the equinoxes. What happens is that the seasons gradually change, and winter becomes summer, and summer gradually becomes winter over the precession cycle. This means that if one were to go back to 10,000 B.C., winter would occur in July and summer in January in the Northern Hemisphere.

Milankovitch did not live to see his theory of global climate change and Earth orbit accepted. It was not until the 1970s that acceptance was gained. The primary reason for this is because Milankovitch did not publish his theories in the major scientific languages of English, French, or German. Today, his theory of global climate change is accepted with modification.

The modification pertains to variations in the Sun's energy output. Radiation from the Sun that is received on the top of Earth's atmosphere is called insolation. Insolation from solar output varies up or down over time.

The Milankovitch Cycles are seen today as a major background factor influencing the Earth's global climate change over long periods of time. The cycles, along with variations in insolation appear to cause a "break point" at which global climate changes. However, Milankovitch Cycles occur within a period of time that is too slow to cause the rapid global climate change that climatologists discovered the Earth periodically experiences.

Milankovitch Cycles suggest the possibility that factors other than greenhouse gases cause global warming. A variety of factors were already discussed here. We have looked at the argument for global warming and what many predict will happen if global warming occurs. We have looked at the argument against global warming, and we have looked at a discussion of factors that might be masking the event of global warming, if in fact, global warming is occurring. It is time to begin exploring the possibility that other factors are influencing the Earth's global climate.

All energy the Earth receives comes from the Sun. In addition to light and heat, the Sun also emits a stream of charged particles called electrons and protons. These charged particles leave the Sun in the form of a solar flare, and they can travel at speeds approaching 3 million m.p.h.

These charged particles from the Sun are called the "solar wind." When the charged particles of the solar wind are ejected from the Sun at a high rate of energy in the form of a solar flare, they can have a dramatic effect on the Earth. Those effects range from power line surges and radio interference to a beautiful display of the aurora borealis. Solar flares most often occur during increased sunspot activity.

We know a sunspot cycle is occurring in one of two ways. The first way is simply to count the number of sunspots. An increase in sunspots indicates an increase in sunspot activity.

The second method is by way of the "Butterfly Diagram." At the beginning of a new sunspot cycle, sunspots mainly appear close to the Sun's north and south poles. As the appearance of sunspots increase, some begin to migrate towards the Sun's equator. This creates a plot of sunspots on the Sun's surface that looks like a butterfly. It is now considered a classic indicator of a new period of sunspot activity.

Sunspots are a unique feature of the Sun. They occupy a region of the Sun where the temperature is cooler than the surrounding surface temperatures. They look dark when compared to surrounding regions. However, sunspot activity over the years is not consistent. There was a period of very low sunspot activity in the latter half of the 17th Century called the Maunder Minimum, after Edward Maunder (1851-1928), one of the first modern astronomers to study the long-term cycles of sunspots. The Maunder Minimum coincided with an abnormally cold period in northern Europe referred to as the Little Ice Age. This fact caused intense investigation in recent years into the influence of sunspots on the Earth's global temperature.

Records of naked-eye sunspot observations in China go back to at least 28 B.C. In the West, the record is somewhat less well documented. It is possible that the Greek philosopher Anaxagora observed a sunspot in 467 B.C. Other possible sunspot observations are scattered throughout ancient literature. A very large sunspot was observed for eight days in 897, but it was dismissed at the time as a passage of Mercury in front of the Sun.

The scientific study of sunspot activity began in the West after the application of the telescope to astronomy in 1609. There is some controversy as to the western astronomer who first observed a sunspot with a telescope. We know that both Galileo and Thomas Harriot observed sunspots toward the end of the year 1610. Johannes Fabricus, David Fabricus, and Christopher Scheiner observed sunspots in March 1611. However, it was Joannes Fabricus who first published on sunspots in his book, *De Maculis in Sole Observatis (On the Spots Observed in the Sun)*, in the autumn of 1611.

After 1645, sunspot activity became a rare event for almost 150 years. When, in 1671, a prominent sunspot was reported, it was treated as a rare event. This changed around 1710 when sunspot activity increased.

Chapter 16
Alternate Truths

Modern studies of sunspots originated with the rise of astrophysics around the turn of the 20th Century. The early chief investigator of sunspots was George Ellery Hale (1868-1938). Hale built the first spectra-heliograph to study the Sun's radiation. Hale was also involved in the building of the Yerkes and Mt. Wilson Observatories, and he was instrumental in the design and building of the 200-inch telescope on Palomar Mountain.

Modern studies, and study of historical records, have revealed some interesting facts about the occurrence of sunspots and the Earth's global climate. The historical record reveals that there were long and cold winters for a 70-year period from about 1645 to 1715. This was during the time that meteorologists call the Little Ice Age. It was a time when the glaciers in the Alps advanced and the North Sea froze over. Northern Lights were almost never seen, and some people considered them to be a myth.

An examination of astronomical records reports almost no observation of sunspots during this period. Sunspots normally increase and decrease on an average 11-year cycle, varying between 7 and 16 years. During the Little Ice Age in northern Europe, this cycle apparently did not occur. The relationship between sunspots and the climate of Earth was confirmed by the study of other sunspot minimums, each of which caused cold weather.

This caused some scientists to investigate if there is a correlation between sunspots and the Earth's global climate, particularly the warming of the Earth's global climate. The observation of sunspots with telescopes and the development of the thermometer occurred at about the same time in Europe. This made it possible for some comparison of sunspot frequency to atmospheric temperature in Europe over several hundreds of years. Scientists are confident they have a reliable record of sunspot activity and global temperature period back to 1500.

A comparison of sunspot activity with Northern Hemisphere air temperature over the 500-year period shows a high correlation between sunspot activity and the Earth's global temperature. The Earth's atmosphere cools when sunspot activity is low, and the Earth's atmosphere warms when sunspot activity intensifies.

This correlation was confirmed by an examination of weather observations between 1580 and 1920 in China's Middle and Lower Yangtze River Valley. The Chinese kept accurate records for the last day of snowfall for each year. It is possible from this Chinese record to determine cooling and warming trends in comparison to sunspot activity.

It was discovered in 1944 that in addition to the 11-year cycle of sunspot activity, the intensity (magnitude) of the 11-year cycle sunspot activity increases and decreases within a 70- to 90-year cycle. This 70- to 90-year cycle is called the Gleissber Period. The Gleissber Period is correlated with increases and decreases in global mean temperature over the last 400 years without any obvious influence from human activity.

Careful statistical studies of temperature variation and sunspot activity come to us from scientists in Denmark. Those scientists determined that there is an association between sunspot activity and long-term variations in global temperature. They were able to demonstrate the association (correlation) to a level of confidence of less than 15%. Said another way, there is an 85% probability that there is an association between sunspot activity and global temperature change. That is a high correlation.

This conclusion throws all computer, climate modeling into disarray. It casts doubts on the reliability of computer models that base their projections of climate change on greenhouse gases. It has long been noted that actual observations of global climate change are inconsistent with model forecasts. The fundamental assumption of carbon dioxide as a force in the greenhouse effect may be incorrect in light of the high correlation between sunspots and global warming.

There is another theory to explain the high correlation between sunspot activity and changes in the Earth's global temperature. This new theory links global temperature change to high-energy particles from the Sun and from space in the form of cosmic rays. Recent satellite data discovered that there is a strong correlation between the Earth's cloud cover and variations in the intensity of cosmic rays. Cosmic rays are energetic atomic particles. When cosmic rays enter the Earth's atmosphere, they ionize (change the electrical charge of) the water vapor and aerosols. This causes the creation of high, thin clouds in the Earth's Stratosphere.

High, thin clouds allow the Sun's energy into the Earth's atmosphere, but they block that energy from reflecting back out into space. This increases the Earth's atmospheric temperature. When the ionization of cosmic rays is high, thin clouds form. The high, thin clouds help retain the Sun's energy in the Earth's atmosphere, causing warming.

In another theory, some scientists suggest that the positions of the continents influence global climate because the configuration of the continents affects the transport of heat by the ocean currents. Plate tectonics is an important process influencing when ice ages occur, and the presence of large land masses at high latitudes appears to be a prerequisite for ice age development. This is because large accumulations of ice cannot form over the oceans.

During the beginning of the ice ages, about 4 million years ago, there were several large landmasses at high latitudes. They included Antarctica, North America and much of Eurasia. This continental configuration led to the creation of ice age conditions.

Plate tectonics contributed to the development of the ice ages in a more subtle way. Plate movements cause the uplift of large continental blocks. Such major uplifts can cause profound changes in global oceanic and atmospheric circulation patterns. In turn, such a shift in oceanic and atmospheric circulation patterns lead to a change of global climate.

As long as the continent of Antarctica exists at Earth's southern pole, the Earth will most likely be pulled back into another ice age. Antarctica was once located near the Earth's equator. Over time, it moved by continental drift to its present location at the South Pole. Once established, continental polar ice caps act like huge cold sinks, taking over the climate and growing bigger during periods of reduced solar output.

Part of the problem with shaking off an ice age is that once ice caps are established they cause solar radiation to be reflected back into space. This perpetuates global cooling. The cooling increases the size of the ice caps, which results in the reflection of even more light from the Sun. However, like the Milankovitch Cycles, continental drift is too slow to explain rapid global climate change.

There is another explanation, and that is a change in the pattern of the Earth's ocean currents. Possibly, the best known of the Earth's ocean currents is the Gulf Stream. The Gulf Stream carries warm tropical waters to the cold Arctic waters off Greenland, Iceland and Norway. There, the prevailing winds carry the warmth brought north by the Gulf Stream across northern Europe. It is estimated that at least 30% of the warmth brought to northern Europe is carried there through this process.

These currents did not always exist. They were created when plate tectonics created the Isthmus of Panama and cut off the circulation of ocean waters between the Atlantic and Pacific Oceans. This had a profound affect on the Earth's climate. The land barrier of the Isthmus of Panama caused the creation of the Atlantic Ocean currents. The Atlantic Ocean current, or the absence of those currents, is what drives some of the Earth's global climate.

Further, certain atmospheric conditions can disrupt this transfer of tropical warmth to northern Europe. An atmospheric low over the Atlantic Ocean will make the waters of the Gulf Stream diverge to the outer rim of the Atlantic Basin region. An atmospheric high over the Atlantic Ocean will cause currents to converge in the middle of the Atlantic. Generally, disruptions to Atlantic Ocean currents caused by an atmospheric low are due to hurricanes. However, if a condition were to occur that created a relatively permanent high over the Atlantic Ocean, the Atlantic currents would be significantly disrupted causing a severe climate change over northern Europe and the northeastern part of North America.

I have intentionally avoided a detailed discussion of ocean currents because I will to come back to them a bit later on. What I do wish to emphasize is that there are strong and significant alternate possibilities to explain global warming. The one I consider most significant is increased radiation from the Sun during intensified sunspot activity. The last maximum of the Gleissber Period occurred in 1932, a period of increased global temperature and global drought. We are now 70 years later at the approach of the next maximum of the Gleissber Period. It is possible that the current trend observed in global warming is not due to greenhouse gases but to increased radiation from sunspot activity.

Whatever the reason, all evidence available states global climate change will occur, and the change will be rapid. There is a reason why scientists know this. The knowledge comes from the ice of Greenland and Antarctica.

Chapter 17
Time Machines

In the past, we thought of time machines in terms of science fiction. This is no longer so. True time machines do exist. They are called ice cores.

Scientists collect ice cores by drilling a hollow tube into a glacier. The stratigraphy of the glacier is then studied. Stratigraphy is the scientific way of saying the ice of glaciers has layers, or strata. The youngest layers of the ice are on the top of the glaciers, and the oldest layers are at the bottom. There are layers in glacier ice because the snow that turns into ice is accumulated more in the winter than in the summer. The annual layers of glacier ice are recognizable. The layers can be counted to work out the age of any part of the glacier ice brought up by an ice core.

The deepest glaciers, (the glaciers with the oldest ice on the bottom) are in the Antarctic and Greenland. The ice cores taken from the Antarctic and Greenland are true time machines that reveal the condition of Earth's climate in the distant past. In Greenland in 1993, the Greenland Ice Sheet Project Two (GISP2) and the Greenland Ice Core Program (GRIP) both obtained ice cores taken from depths of over 9,000 feet. The age of the ice obtained from the bottom of these ice cores was over 110,000 years old. An ice core drilled by French and Soviet scientists at an Antarctic site called Vostok obtained an ice core record at a depth of almost 10,000 feet. The oldest ice brought up by that ice core is over 420,000 years old, extending into the past over more than three ice ages.

The detail of the Earth's climate revealed by ice cores is remarkable. This is because each layer of ice in an ice core represents a single year, and almost everything that fell in the snow during that year remains in the ice. This includes wind-blown dust, ash, and atmospheric gases.

The most important information to the study of the Earth's past climate lies in the small bubbles of atmospheric gases trapped in the ice. The bubbles of atmospheric gas are actual samples of ancient air. The gas bubbles can tell us exactly what the composition of the Earth's atmosphere was when the gas was trapped in the ice.

The second piece of important information that can be derived from ice cores is indirect, derived from the oxygen atoms that form the water molecules in the ice. All atoms are composed of protons, neutrons, and electrons. The total number of protons and neutrons provide the atomic weight of any atom.

Oxygen has an atomic weight of 16, consisting of 8 protons and 8 neutrons. However, some oxygen atoms have an extra neutron, and thus, they are heavier. About 1% of all oxygen atoms have an atomic weight of 18. In a sample of glacial ice, the ratio between oxygen atoms with a normal atomic weight of 16 and the heavier oxygen atom with an atomic weight of 18 is determined by air temperature.

If the temperature is cold, a larger proportion of normal oxygen atoms are present. If the air is warmer, there are a proportionally larger number of the heavier oxygen atoms. By measuring the ratio between these two different forms of oxygen atoms, an approximation of the temperature of Earth's climate can be made from the ice core sample.

Ice core records provide a direct, detailed and complete record of past global climate change. This makes ice core records valuable for comparison of what is happening with the Earth's climate today to past climates. Ice core records document a wide range of environmental parameters that are responses to climate change. Examples of information about the Earth's global climate that can be derived from ice core samples are: composition of atmospheric gases, atmospheric circulation, global temperature, and global precipitation.

Ice core records also document many of the claimed causes of climate change such as solar variability, volcanic activity, and the presence of greenhouse gases. Ice core records have a high resolution of accuracy, and they cover a large span of time. They also provide us with a closer look at what occurred during the ice ages that have dominated the Earth's climate for at least 4 million years.

We think of ice caps on Earth as normal because in the recent past there have always been ice caps. However, there were periods in the Earth's history when there were no polar ice caps. Yet there were many periods in the Earth's history when ice age climates were dominant and large parts of the Earth were covered by glacial ice. Ice cores reveal that the Earth does not have a stable climate.

Ice ages begin when a continental ice sheet forms at high latitudes and the winter accumulation of snow fails to completely melt in the summer. More snow is added in subsequent years, and it is the weight of the snow compacting the lower layers of snow that turn the older snow into glacial ice. The weight of the snow and ice is what causes the glacier to flow.

The uppermost part of a glacier is called the "zone of accumulation." In this location, there is frequent snowfall and little melting, even in summer. The toe of the glacier is known as the "zone of ablation." Here the ice melts, debris is deposited, and rivers derived from the glacial melt water are formed. A glacier is always in a state of balance, or "dynamic equilibrium." If there is more accumulation than ablation, the glacier will advance. If there is more ablation than accumulation, the glacier will recede.

Rocks that are pushed at the front of a glacier are called "moraines." At the toe of a glacier that has stopped advancing there will be a terminal moraine. A terminal moraine is how we know how far south the glaciers of the last Ice Age advanced.

The last 130,000 years saw the Earth's global climate system change from a warm interglacial period to an ice age and back again to a warm interglacial period. An interglacial is a period of warm weather between ice ages. The period in which we are currently living is one such interglacial period, a geological epoch called the Holocene.

Another interglacial period occurred between 110,000 and 130,000 years ago and lasted about 20,000 years. It is called the Eemian Interglacial. The Eemian Interglacial occurred during the Pleistocene Epoch.

The Eemian Interglacial was a time of global warming between ice ages, much like the climate we are experiencing today. The Eemian Interglacial is seen as a close counterpart to the interglacial period in which we now live. The Earth's climate was about as warm, and precipitation was about the same as it is today. Mean sea level was similar. Accepting that there is a general similarity between these two warm periods, the Eemian Interglacial is used to predict the duration of global warming in which we currently live. It is also used to study the possibility of sudden climate variability during our time.

There are indications of large-scale climate instability in the middle of the Eemian Interglacial, and since we are in the middle of a similar interglacial period, this has generated much interest. If dramatic climate changes occurred during the Eemian Interglacial, such changes could also occur now. There is evidence that several times during the warming of the Eemian Interglacial the Earth's global climate shifted abruptly back to a colder climate.

The evidence from ice cores suggest the climate changes during the Eemian occurred within a 10-year period of time. When the Earth's global climate cooled during the Eemian Interglacial, it cooled by about 4.5° F to 7° F., and the change to a cold climate lasted for several hundred years. These events occurred in sudden steps. They demonstrate the recent finding that the Earth's climate tends to remain stable over many years, then, it abruptly changes within a few short years.

The Greenland and Antarctica ice cores made significant contributions to our understanding of the Earth's climate during the Eemian Interglacial. However, a significant portion of our knowledge about the Earth's climate also comes from cores drilled into ocean and lake sediments. Oxygen isotopes recorded in the shells of tiny ocean creatures called foraminifers, a primary component of ocean sediment, provide a measurement of global climate change, just as do the oxygen isotopes in ice cores. Plant pollen deposited in northern European lake sediment provides another record of the transition on land between woodlands that are prevalent during warm weather and the grasslands that are prevalent during cool weather.

What interests, and concerns, scientists about the Eemian Interglacial is that all current data suggests there were periods when the Earth's global climate rapidly cooled. It was at the end of the Eemian Interglacial that the Earth slipped back into the last Ice Age. The problem is scientists do not know why the last Ice Age occurred.

For a possible answer, we can look closer to our own time. A striking occurrence of a fast change in global temperature occurred about 12,000 years ago, when the Earth was emerging from the last Ice Age. The gradual warming across the Earth was interrupted by a sudden return to ice age conditions in an episode called the Younger-Dryas Event. Many are now convinced that a large influx of fresh water into the Earth's oceans caused a change in the flow of ocean currents. That in turn caused rapid cooling of the Earth's global climate during the Younger-Dryas Event.

Chapter 18
Gyres

The Sun, the wind, and the spinning of the Earth all work to keep the oceans of the Earth in motion. This was so from the beginning of Earth's geological history, and it is so today. The Earth's winds and rotational forces cause the water of oceans to spin to the right, or clockwise, in the Northern Hemisphere and to the left, or counterclockwise, in the Southern Hemisphere. Easterly trade winds near the Equator push ocean waters to the west, while westerly winds in the mid-latitudes push ocean waters to the east.

The wind and the spin of the Earth's rotation combine forces to create huge circulating ocean currents called gyres (from the Greek *gyros*, meaning "circle"). Both the Earth and its rotation are spherical. This results in narrow, swift, western boundary ocean currents. Examples of such currents are the Gulf Stream in the Atlantic and the Kuroshio and Indigo Blue Currents in the Pacific Ocean. The Earth's giant spinning ocean gyres are linked to a longer, deeper water journey. The main part of this global ocean circulation occurs in the northern Atlantic Ocean, where the cooled water becomes so heavy that it plunges downward, displacing the water below. This cold, heavy water travels southward at great depths and it eventually spreads around the Antarctic circumpolar regions in an ocean current system know as the "Atlantic Conveyor."

In the Antarctic, the water mixes with the warmer water from regions of the Indian and Pacific Oceans. It then reenters the Atlantic Ocean, where, as warmer water, it rises to the surface and flows back northward again to Greenland and the Labrador Sea. There it once again becomes cold, heavy water, starting a new cycle.

The grand voyage of the Atlantic Ocean's waters is called a thermohaline circulation (thermo for heat and haline for salt). The thermohaline circulation pattern of the Atlantic Ocean works like this. The Gulf Stream carries warm and relatively salt-free surface water from the

Gulf of Mexico northward to the seas between Greenland, Iceland and Norway. Once in the northern latitudes, westerly winds sap the heat from the seawater. The winds increase the salinity (amount of salt content) of this seawater through evaporation. This cold, salty water has a density greater than the water surrounding it, and it sinks in several underwater waterfalls with a volume greater that 30 times the output of the Amazon River. The waterfalls are called "thermohaline sinks."

The sinking cold, salty water pulls in more new water behind it. This is what causes the Atlantic Ocean circulation that warms both northern Europe and northern portions of the North American continent. The sinking of this cold, salty seawater will not occur if the water loses its salt content.

Fresh water floats on top of salt water. If there is a sudden, massive introduction of fresh water into the areas around Greenland and Iceland where the underwater waterfalls that drive the Atlantic currents occur, those currents will either stop or shift sharply to the south. What concerns many is that this event may already be well underway. Rainfall, which is fresh water, has increased over the North Atlantic. Both the Arctic Ice Cap and the great Greenland glaciers are melting. The fresh water from the melt of this ice is being introduced into precisely the area of the Atlantic Ocean where the thermohaline sinks occur.

To add to the concern, recent observations of the salinity of ocean water in areas critical to the thermohaline sinks show the salinity of the water is decreasing. It is feared that a freshwater lens is in the process of forming over the salty seawater. If that occurs, northern Europe is in store for some cold winters.

This is critically important to an understanding of global climate change presented here. This system of ocean flow is called the Atlantic Conveyor, and it is a system of thermohaline-driven ocean currents that will be disrupted when fresh water is introduced into it.

The Gulf Stream circulates clockwise around the North Atlantic Ocean. In its clockwise circulation around the North Atlantic Ocean, the Gulf Stream picks up warm tropical water in the Caribbean, and it carries that warm tropical water northward to Greenland and Iceland. An interesting sequence of events happens when the Gulf Stream (and other similar thermohaline currents) reaches the northern waters of the Atlantic Ocean.

The northern Atlantic Ocean south of Greenland and Iceland is an area of the Earth dominated by westerly winds. Westerly winds are winds that blow from west to east. The relative warmth and moisture of the Gulf Stream water is transferred to the westerly winds, and the westerly winds in turn transfer the warmth and moisture picked up from the Gulf Stream to the land areas of northern Europe.

The westerly winds have a profound effect on the waters in the Gulf Stream. The westerly winds cause two things to happen in the Gulf Stream. First, the water becomes colder from the loss of heat transferred to the wind. Second, the wind evaporates the water, and the remaining water becomes saltier.

Cold, salt water is dense, and this dense water sinks to the bottom. It sinks, and it sinks rapidly. The sinking is so rapid that it creates several great underwater waterfalls. The most significant of these underwater waterfalls is just off the southern tip of Greenland. This sinking water is important, because as it sinks, it creates a vacuum that pulls in more warm water from the south. This is what keeps the cycle going and the current flowing

This cycle can be disrupted by the introduction of large amounts of fresh water into the area of the underwater waterfalls. Fresh water floats on top of salt water in what is called a freshwater lens. If a large amount of fresh water is introduced into the Atlantic Ocean south of Greenland and Iceland, the fresh water will float on top of the warm salty water brought up from the tropics by the Gulf Stream.

In this event, the westerly winds will not pick up the warmth of the Gulf Stream and transfer that warmth to northern Europe. The freshwater lens will also prevent evaporation of the warmer, saltier Gulf Stream water. The result will be the loss of the thermohaline sinks that draw in new water. The Gulf Stream stops, and the climate of northern Europe returns to ice age conditions.

The situation is further aggravated because fresh water freezes faster than saltwater. Arctic ice sheets and icebergs will spread south. Because ice is an efficient reflector of sunlight, the cooling of the climate accelerates at an increased rate.

Abundant fresh water exists to make this scenario happen. The fresh water is currently contained in the ice sheets of the Arctic and the great glaciers of Greenland, both of which are directly north of the Gulf Stream's thermohaline underwater waterfalls. The fear is that continued global warming will cause this ice to melt and flow into the North Atlantic Ocean. It is also feared that global warming will increase rainfall in this region. Both events will introduce large quantities of fresh water into the North Atlantic Ocean at the location where the thermohaline sinks occur. This will cause a shutdown of the ocean currents that warm Europe and northeastern North America.

If these events occur, they will occur rapidly. One only need take a look at the winter freezing of sea ice in Antarctica at the approach of winter to see just how fast such events occur. At the onset of winter in Antarctica, the sea ice freezes at 40,000 square miles per day. If the disruption of the Atlantic thermohaline sinks occurs, it will create climatic conditions similar to those found in Antarctica in the winter. There is no reason to believe that the spread of sea ice will not happen with a similar speed.

What is frightening is that our great time machine, the Greenland ice cores, tells us such an event has happened at least once approximately 12,000 years ago, and possibly several times over the last 110,000 years ago. The irony is that global warming will cause a melting of ice that will create the fresh water that stops the thermohaline sinks, causing global cooling.

This return to a cold Earth has happened before, and it will happen again. It will not occur gradually over hundreds of years. All evidence we have today suggests that the switch will occur within 4 to 10 years. So we can see that while our time machines reveal interesting information, sometimes a time machine will reveal information we do not want to see.

Chapter 19
It's Happened Before

The Bishop of Geneva was called to Chamonix at the foot of Mont Blanc in 1645 to do a Rite of Exorcism. The Mer de Glace Glacier (Sea of Ice Glacier) was slowly flowing over farms and villages. Glaciers were advancing throughout Europe during the 1600s and 1700s, and they were progressing down the mountains farther than they had for thousands of years. Sea ice in the North Atlantic Ocean destroyed the Scandinavia and Iceland fisheries, and in China, severe winter conditions killed orange groves that thrived for centuries in Jiang-Xi Province.

The Bishop of Geneva did the Rite of Exorcism, and the ice retreated. But it was not long until the ice returned, and the Bishop of Geneva was again called in to a struggle against ice that would last for years. It was the time of the Little Ice Age.

Europe went through a period of unprecedented warming between 1000 and 1300, a period called the Medieval Warm Period. During this time, the Earth's climate warmed, and temperatures were on the average 3° F to 5° F warmer than they are today. The warm weather led to better crops that supported a large population. The prosperity encouraged the building of cathedrals and the beginning of the Age of European Exploration.

Sea ice off the coast of Iceland nearly vanished for 3 centuries. Eskimos settled Ellemere Island at the usually frigid northwest corner of Greenland. The Vikings colonized southern Greenland. A warming trend occurred in Alaska, and the warmth pushed the snow line of the Rocky Mountains about 1,000 feet higher than where it stands today. Then, it got terribly cold. Sometime between 1250 and 1350, a chill in the Earth's global temperature set in, dropping the temperature on average to 1.5° F to 3° F colder than it is today.

The Norse settlements in Greenland and Iceland were the first

European communities affected by the cooling climate. The Greenland colony failed sometime between 1270 and 1370. The Iceland Colony barely survived when, after a loss of over half its population, the Icelanders turned to fishing for food.

The height of the Little Ice Age occurred between 1550 and 1750. Glaciers in Europe advanced over farms, villages, and valleys. The Baltic Sea and the Thames River regularly froze over. Frost Fairs on top of the Thames River became common in the 1700s, and in 1820 it was reported that the Thames River froze over to a depth of 5 feet. By 1700, the expanse of sea ice around Iceland was such that the open ocean could not be seen from the highest point on the island during the winter.

Crops failed in Europe. Famine, disease and warfare raged across the continent. Infanticide and abortion increased as families ran out of food. Timberlines retreated down mountains, and agricultural areas were abandoned to the cold. Eskimos driven out by the ice, paddled as far south as Scotland. According to some research, the rates of famine and mortality increased all over the world as a result of the Little Ice Age. Diseases such as Bubonic Plague and anthrax took advantage of a human population weakened by starvation.

Native American tribes such as the Iroquois relocated their villages to escape the cold. These migrations stirred up political conflicts among the tribes, leading to the creation of the Iroquois League sometime around 1550.

Further, there is evidence that the glaciers of South Patagonia, South America, advanced during the 17th, 18th, and 19th Centuries. All totaled, it is estimated that the Little Ice Age—through the direct effects of a cooling global climate and through the indirect effects of disease, starvation, and warfare—resulted in the loss of 40% to 60% of all human population worldwide.

It is generally accepted that the Little Ice Age did not end until about 1900. However, a recent calibration of the dating for several ice cores drilled in Wyoming indicates that the Little Ice Age ended as abruptly as it began (within a period of 4 years), sometime around 1870. But it continued to influence global climate until about 1914.

The implication of this is that the global warming we are seeing today is simply a return to the period of warmth that existed during the Medieval Warm Period. Global warming occurring today is a rebound from the Little Ice Age. It is not the result of human-introduced greenhouse gases.

Earth's Future Climate | 103

There is debate over exactly what caused the Little Ice Age. As discussed in earlier chapters, the Sun is the ultimate source of all the Earth's warmth. Many researchers look to the Sun for an answer to the question of global climate change. In the 1970s, several researchers noted a correlation between sunspot activity and the Earth's global temperature. It was found that a period of low sunspot activity, named the Maunder Minimum, which occurred between 1645 and 1715, matched the coldest period of the Little Ice Age.

Another possibility, other than a fluctuation of the Sun's output, is that a deterioration of the Gulf Stream in the Atlantic Ocean caused the cool global climate of the Little Ice Age. The Medieval Warm Period was a time when global temperature averaged 3° F to 5° F warmer than today. It is possible that this warming brought about increased precipitation and a melting of Arctic ice that interrupted the Atlantic Ocean's thermohaline sinks.

The theories for the cause of the Little Ice Age are in dispute. However, if we step back 12,000 years in Earth's history, we can see a precise event of global cooling that was caused by global warming. That event is called the Younger-Dryas.

The last Ice Age began to thaw about 15,000 years ago. Five thousand years later, the great ice sheets were in retreat towards the North Pole, and the Earth was about as warm as it is today. This escape from the cold was abruptly interrupted by the Younger-Dryas Event.

The Younger-Dryas Event was a significant, rapid climate change. During the Younger-Dryas Event, a global climate that was in a state of rapid warming suddenly reversed itself. It cooled to an average temperature that was between 5° F and 7° F colder than global temperatures today. The ice cores tell us the change occurred within 4 to 10 years, and the cold climate lasted for approximately 400 years.

What caused the Younger-Dryas Event? The event appeared at the time when the Earth was warming. We think we know the answer.

When the ice of the last Ice Age began to melt, the melt water formed two huge lakes. One lake, the Great Siberian Lake, formed in western Siberia. The other lake formed in Canada, and is called Agassis Lake. Both of these were freshwater lakes, and they were blocked from releasing their water to the ocean by ice dams that were the remnants of the glaciers from the last Ice Age.

Eventually, a warming climate eroded the ice dams, and they broke. This released the fresh water from both of these large lakes into the Arctic

Ocean. The Arctic Ocean outlets into the Atlantic Ocean through the Labrador and Norwegian Seas, both of which are situated directly over the thermohaline sinks of the Gulf Stream.

It was this huge pulse of fresh water that created the Younger-Dryas Event. The massive introduction of fresh water disrupted the thermohaline currents of the Atlantic Ocean by forming a fresh water lens, stopping the sinking of water that drives the Gulf Stream. The warmth brought north by the Atlantic currents from the tropics stopped. The Earth turned cold.

Analyses of ocean sediments in the tropical waters of the western North Atlantic Ocean demonstrate that these waters were warm when the northern waters were cold during the Younger-Dryas Event. This in turn demonstrates that the thermohaline circulation of the Gulf Stream shut down during a time when tropical oceans were retaining their heat. This was determined from cores of ocean floor sediments that measure the amount of organic matter falling to the ocean's floor. More organic matter grows in warmer temperatures. Therefore, the more organic matter there is in ocean sediments, the warmer was the temperature at the time the sediments were laid down.

This observation of increased organic material in ocean sediments during the Younger-Dryas Event is found only in the tropical waters. If greenhouse gases were responsible for the cooling of Earth's global climate during the Younger-Dryas Event, the influence of the cooling would have appeared everywhere on the Earth at about the same time. This is not the case for the Younger-Dryas Event, and it supports a disruption of the Gulf Stream's thermohaline circulation as the cause for the cooling.

There is no way to escape the conclusion that the Earth's climate can, and does, change abruptly and often. The Greenland ice cores tell us that the cooling of the Younger-Dryas Event both started and ended within a period of 10 years. The same Greenland ice cores tell us that similar, abrupt changes in global climate occurred 110,000 years ago during the Eemian Interglacial.

There are no longer any great post-Ice Age lakes to release millions of cubic miles of fresh water into the oceans of the world. Yet that water still exists in the ice of the Arctic Ocean and the Greenland Glaciers. It will have an influence on what happens if global warming occurs. Global warming will increase rainfall.

Much of the rain in the Northern Hemisphere will fall on the North

Atlantic Ocean at the place where the Gulf Stream's thermohaline sinks occur. Global warming will melt both the ice pack of the Arctic Ocean and the great glaciers of Greenland. Both bodies of ice can put enough fresh water into the Atlantic Ocean thermohaline sinks to stop those sinks. It has happened before. It will happen again. Global warming will cause global cooling.

Chapter 20
Ockham's Razor

Ockham's Razor states that the simple explanation is usually the best. In scientific terms, Ockham's Razor is parsimony. The Principle of Parsimony states that when presented with many complex solutions to a problem, the solution that is least complex is the solution accepted.

The question of global climate change presents several complex arguments. There are some who state that there is a correlation between the emission of greenhouse gases, such as carbon dioxide, and global warming. We know from the Snowball Earth Theory that carbon dioxide played a major role in keeping the Earth from becoming a frozen ice ball between 750 and 550 million years ago.

However, with the introduction of organic life during the Cambrian Explosion, the Earth's levels of greenhouse gases, particularly carbon dioxide, have been regulated by that organic life for at least the last 500 million years. As a result, we do not know today if an increase in carbon dioxide in the Earth's atmosphere is causing global warming, or if an increase in global warming is causing an increase in carbon dioxide due to increased biological activity.

This question is further complicated by the fact that water vapor is the most powerful of the greenhouse gases, somewhere between 3 to 5 times more powerful than carbon dioxide. Water vapor manifests itself in the Earth's atmosphere as clouds. Some clouds act to increase global warming while other clouds act to cool the Earth's atmospheric temperature. Cloud cover, and therefore, the amount of water vapor in the Earth's atmosphere, is projected to increase if global warming increases the atmospheric hydraulic cycle of evaporation. We simply do not know what the effect of increased water vapor in the Earth's atmosphere will be.

There are also problems with the location of the surface-based instruments that measure the Earth's atmospheric temperature. Many of these instruments are located in "heat islands"—major metropolitan areas and airports where the presence of concrete and asphalt cause artificially high readings. Even instruments in rural areas record artificially high readings if trees that surround them are too close. This creates a bias in favor of global warming.

There are too few surface-based instruments to provide an accurate view of global climate change, and there is a concentration of surface-based instruments in North America. As a result, the atmospheric temperature readings in North America have a greater influence on global temperature change than readings from the rest of the world. For example, central and northeastern Asia, the Himalayan Mountains, the Andes Mountains, sub-Saharan, central Africa, and Antarctica have few recording stations for huge areas of land. Further, the Earth's oceans cover approximately 70% of the planet's surface, and there are fewer than 1,000 buoys recording atmospheric temperature over the Earth's oceans.

Satellites overcome this problem of coverage, providing readings of the Earth's atmospheric temperature over 97% of the planet's surface on a daily basis. Satellites have preformed these observations since the early 1970s. However, there were initially problems with the calibration of satellite atmospheric temperature readings. The early problems with calibration are now corrected.

If you accept that satellite information is now correct, then you must also accept what that information reveals. Satellite measurement of the Earth's atmospheric temperature reveals a slight cooling of the Earth's global temperature by about 0.5° F since the early 1970s. This cooling was also confirmed from measurements made by weather balloons.

There are many ongoing efforts to create a computer model that will project future changes in global climate. Computer models are influenced by the GIGO Principle, "Garbage in, garbage out." Computer models are only as accurate as the information programmed into them. As seen above, current measurement of the global climate change is subject to bias. That is to say, computer models are dependent upon instrument readings that may not be accurate. If computer models are not receiving accurate information, they cannot provide an accurate projection of future climate changes.

There are other sources that might be causing climate change, but those sources are not El Niño and ozone holes. El Niño and ozone holes are amplifiers. They enhance (or increase) the effect of global climate change put in place by other mechanisms, but of themselves, they are not the cause of global climate change.

Volcanoes are another aspect of the Earth's environment that can cause a change in atmospheric temperature. The Snowball Earth theory suggests that volcanoes have ejected enough carbon dioxide into the Earth's atmosphere to create a warm climate. When an abundance of biological life appeared on the Earth during the Cambrian Explosion approximately 500 million years ago, that biological life acted to control the amount of carbon dioxide volcanoes ejected. That process is still in place today, and there is no evidence to suggest that carbon dioxide ejected from volcanoes influence a long-term change in global temperatures.

Aerosols ejected from volcanoes are another story. Those aerosols act to cool the Earth's global temperature by reflecting sunlight away from the Earth. The eruption of the Tambora Volcano in 1815 ejected enough aerosols to stop the advent of summer in North America for several years, and the eruption of Toba 70,000 years ago almost resulted in the extinction of *homo sapiens*.

Aerosols created by industrial pollution and wind-blown dust have a similar effect. All result in a cooling of global atmospheric temperature, and they are, therefore, not mechanisms for global warming.

A change in the Earth's ocean currents can cause significant global climate change. It is important to understand that, in the short term, in the geological scale of time, ocean currents do not change climate. Rather, it is a change in climate that changes ocean currents. However, once the ocean currents do change, they influence global climate.

Tectonic changes in the Earth's land surface and the Milankovitch Cycles of the Earth's orbital relationship to the Sun have in Earth's past history, played an important part in global climate change. However, each event occurred slowly—over many hundreds of thousands of years—too slow to explain a rapid change in global temperature discovered by the ice cores. Further, we are right in the middle of the Milankovitch Cycles, which means we are approximately halfway between the minimum and maximum extremes in global climate that those cycles can cause.

Sunspots are another factor. There is clear evidence that sunspot activity, or the absence of sunspot activity, does affect the Earth's global temperature. A correlation coefficient is a mathematical statement of the probability that a cause and an effect are related. The number is always stated as a fraction, and the closer that fraction is to one, the higher the probability that a cause and an effect are related.

There is a meticulously conducted study by two Danish scientists, Svensmark and Friis-Christensen that shows a correlation coefficient of .92 that sunspots have an influence on the Earth's global atmospheric temperature. (Wesker, "Climate Change: A Summary of Some Present Knowledge and Theories", and Svensmark and Friis-Christensen, "Variation of Cosmic Ray Flux and Global Cloud Coverage—a Missing Link in Solar-Climate Relationship.") The Danes' finding of a .92 correlation coefficient between sunspots and global warming means there is a better than an 84.6% probability that sunspots are responsible for changes in the Earth's global atmospheric temperature.

This study concentrated on the Gleissber Period. Gleissber Periods are 70- to 90-year cycles in the intensity of the 11-year sunspot cycle. An 11-year sunspot activity will be more intense at the height of a Gleissber Period than at the bottom of a Gleissber Period. The last peak of the Gleissber Period occurred sometime around 1932. This is the period when most of the Earth's global warming occurred during the 20th Century.

The lack of sunspot activity is believed to have contributed to the cool climate of the Little Ice Age. Sunspots add to the energy the Sun emits. If there is a decrease in sunspots, this means the Sun is emitting less energy. It takes only a 1% decrease in the Sun's energy to cause the Earth's climate to revert back to ice age conditions.

The Danish study looked at Gleissber Periods back to 1500 and compared the climate during Gleissber Periods to western European records of climate change. The Danes are seriously worried about the potential for any increase in sea level due to global warming, because like Holland, a large portion of their country is only a few feet above sea level. They cannot afford to make a mistake in their research. An important aspect of the Danish research is that it projects that we are now entering into a period when the Gleissber Period is approaching its peak, and we should be experiencing a period of global warming.

Other data regarding the Earth's global temperature during the 20th Century appears to support the Danes' projection. The last peak of the Gleissber Period occurred sometime around 1932. This was a period of global warming and drought. Thirty-five years later, at the low point of the Gleissber Period during the late 1960s and early 1970s, global cooling occurred. The cooling was sufficiently significant to cause scientists to publish warnings about the cooling in journal articles. And now, 35 years after that cooling, we are again looking at global warming at the peak of the 70-year cycle of the Gleissber Period, with the accompanying warnings about global warming.

On the other hand, while a 1% decrease in solar radiation can bring back another ice age so could a 1% increase in solar radiation. There is now fairly strong evidence to suggest that the influx of fresh water from melting Artic ice into the northern Atlantic Ocean can bring about a collapse of the great Atlantic conveyor currents. The failure of those currents to bring the warmth of the tropics north has in the past caused Earth's global climate to cool significantly. This is what appears to be happening today. A 1% increase in solar radiation can bring about the warming of Earth's climate to cause that to happen.

Since the development of an atmosphere, the Earth has protected life on its surface from the deadly radiation of the Sun. We know that the intensity of the radiation from the Sun varies over cycles, and there is strong evidence to suggest that variation in the radiation of the Sun affects the Earth's global temperature. What we do not know is what effect our tampering with the content of the Earth's atmosphere will have. The complexity of the question makes it currently beyond our ability to answer.

The conclusion here is that global warming is caused by the increased sunspot intensity of the Gleissber Period. It is the simplest and least complex explanation. Therefore, the Principle of Parsimony states that this is the explanation that must be accepted in the absence of a preponderance of proof for one of the other more complex solutions.

This, in turn, will produce another ice age. What is important to note is that global climate change from a period of warmth to an ice age has happened in the past in less than 4 years. This is the amount of time that stands between the warm global climate we enjoy today and the next ice age.

Our modern civilizations are not organized to resist such a dramatic global climate change. When the change comes, it will be so rapid there will be no time to respond. The Theory of Evolution states that all species will eventually go extinct. Severe climate change carries the potential for our extinction. It is somewhat ironic that extinction is based upon the most humble of elements, salt. As stated many times here, it is upon the most humble among us that the survival of all depends.

Epilogue

There is always a lag between the research and the time a book such as this appears in print. That lag is usually 1-1/2 to 3 years. This book is no exception to that lag time. The research for this book was conducted 1-1/2 to 2 years ago. How much can change in two years?

The United States is now combating an epidemic from the mosquito borne disease the West Nile Valley virus. The United States has the dubious honor of looking forward to other such mosquito borne pathogens such as malaria and yellow fever as global warming continues to allow a northward expansion of these tropical diseases.

There is also no doubt now that global warming is happening. It is seen in the melting of glaciers and in the change of precipitation patterns. There has been no change in annual precipitation, except that the precipitation comes all at one time. That is to say, there are long periods of drought followed by short, but extreme, precipitation events. This is what is predicted by global warming, and it is exactly what is happening.

There is not yet a consensus as to the cause for global warming. As suggested here, the base cause for global warming is an increase in solar activity. However, the activities of man may be providing an amplifying influence to that warming which is causing its effects to be greater than that which would occur under normal conditions.

The unfortunate and tragic events of September 11, 2001 support this conclusion. For several days after that event, all aircraft in the United States were grounded. It was noted here that one of the indicators of global warming is the rise in the minimum nighttime temperature. During the grounding of all aircraft after the September 11th attacks, those nighttime minimum temperatures began to drop and return to their normal levels. What this finding suggests is that the injection of aircraft exhaust into the high stratosphere has a significant influence on global warming. No one previously saw this.

This fact acts to contribute to the complexity of the arguments surrounding global climate change. There can be no doubt that global warming is now occurring. The big question is, what will happen as a result of the warming?

Glossary

Absolute Zero: A point of temperature, theoretically equal to -273.15° C, or -459.67° F, the hypothetical point at which a substance would have no heat or molecular motion.

Absorption of Radiation: The uptake (or absorption) of radiation by a solid body, liquid or gas. The absorbed energy may be transferred or re-emitted.

Acid Rain: Also known as "acid deposition." Acidic aerosols in the atmosphere that are removed from the atmosphere by wet deposition (rain, snow, fog) or dry deposition (particles sticking to vegetation). Acidic aerosols are present in the atmosphere primarily due to discharges of gaseous sulfur oxides (sulfur dioxide) and nitrogen oxides from both anthropogenic and natural sources such as volcanoes. In the atmosphere, these gases combine with water to form acids.

ACOUS: Arctic Climate Observations Using Underwater Sound.

Aerosols: Particles of matter, solid or liquid, larger than a molecule but small enough to remain suspended in the atmosphere. Natural sources include salt particles from sea spray and clay particles as a result of weathering rocks, both of which are carried upward by the wind. Aerosols can also originate from volcanic eruptions or as a result of human activity and in that case are often considered pollutants.

Albedo: The amount of light reflected by a geographic area, such as land, water, and ice. It is a measurement of the ratio of reflected light to incident light. Albedo can be expressed as either a percentage or a fraction of 1, and the higher the number the greater is the amount of light reflected. Snow covered areas have a high albedo (up to about 0.9 or 90%) due to their white cover, while vegetation has a low albedo (generally about 0.1 or 10%) due to the dark color and light absorbed for photosynthesis. Clouds have an intermediate albedo and are the most important contributor to the Earth's albedo. The Earth's aggregate albedo is approximately 0.3.

Alliance of Small Island States (AOSIS): The group of Pacific and Caribbean nations who call for relatively fast action by developed nations to reduce greenhouse gas emissions. The AOSIS countries fear the effects of rising sea levels and increased storm activity predicted to accompany global warming. Its plan is to hold Annex I Parties to a 20 percent reduction in carbon dioxide emissions by the year 2005.

Amplitude: A measurement of the distance between the high point and low point of any event that occurs in a cycle of high's and low's. The greater the difference between the high point and the low point of the cycle, the greater is the amplitude.

Annex Parties: Industrialized countries that, as parties to the Framework Convention on Climate Change, have pledged to reduce their greenhouse gas emissions by year 2000 to 2005. Annex I Parties consist of countries belonging to the Organization for Economic Cooperation and Development (OCED) and countries designated as Economies-in-Transition.

Anoxic Conditions: Water with too little oxygen to support life.

Antarctic Circumpolar Wave: An ocean current wave propelled by the Antarctic Circumpolar Current and atmospheric highs and lows; areas of warmer and cooler water circulate around the continent, taking eight years to complete the journey. During three of those years, warm water moves from South America to southern Africa.

Anthropogenic: Human-caused or derived from human activities.

Aphelion: The point the Earth's orbit (or the orbit of any other stellar body) is most distant from the Sun (or the object around which it orbits).

Atlantic Conveyor: A system of ocean currents that carrying warm water north to Europe and then continue around the world to end in the Marianis Trench in the Pacific Ocean.

Atmosphere: The mixture of gases surrounding the Earth. The Earth's atmosphere consists of about 78.1% nitrogen (by volume), 20.9% oxygen, 0.036% carbon dioxide and trace amounts of other gases. The atmosphere can be divided into a number of layers according to its mixing or chemical characteristics, generally determined by its thermal properties (temperature). The layer nearest the Earth is the troposphere, which reaches up to an altitude of about 5 miles in the Polar Regions and up to nearly 11 miles above the equator. The stratosphere, which reaches an altitude of about 31 miles lies atop the troposphere. The mesosphere, which extends upwards to 50 to 80 miles, is atop the stratosphere, and finally the thermosphere, or ionosphere, gradually diminishes and forms a fuzzy border with outer space. There is relatively little mixing of gases between layers.

Baseline Emissions: The emissions that would occur without policy intervention (in a business-as-usual scenario). Baseline estimates are needed to determine the effectiveness of emissions reduction programs (often-called mitigation strategies).

Berlin Mandate: A ruling negotiated at the first Conference of the Parties (CoP1), which took place in March 1995, concluding that the present commitments under the Framework Convention on Climate Change are not adequate. Under the Framework Convention, developed countries pledged to take measures aimed at returning their greenhouse gas emissions to 1990 levels by the year 2000. The Berlin Mandate establishes a process that would enable the Parties to take appropriate action for the period beyond 2000, including a strengthening of developed country commitments, through the adoption of a protocol or other legal instruments.

Biochemical Oxygen Demand (BOD: The demand that various biological organisms and chemicals place on the oxygen supply within a given body of water.

Biogeochemical Cycle: The chemical interactions that take place in the atmosphere, biosphere, hydrosphere, and geosphere.

Biomass: Organic non-fossil material of biological origin. For example, trees, plants, insects, animals, and the like, are parts of the Earth's biomass.

Biomass Energy: Energy produced by combusting renewable biomass materials such as wood. The carbon dioxide emitted from burning biomass will not increase total atmospheric carbon dioxide if this consumption is done on a sustainable basis (i.e., if in a given period of time, re-growth of biomass takes up as much carbon dioxide as is released from biomass combustion). Biomass energy is often suggested as a replacement for fossil fuel combustion.

Biosphere: The region on land, in the oceans, and in the atmosphere inhabited by living organisms

Borehole: Any exploratory hole drilled into the Earth or ice to gather geophysical data. Climate researchers often take ice core samples, a type of borehole, to predict atmospheric composition in earlier years. Boreholes are also drilled into the sediments on ocean and lake bottoms for the same reason.

B.p.: Before present. A method of measuring time in years before the present year. For example, A.D 500 is written in this method of notation as

1,500 b.p. (1,500 years before A.D. 2000). It also written in some scientific disciplines as y.b.p., years before present.

Butterfly Pattern: A classic pattern of sunspots that indicates the beginning of increased sunspot activity. It is the spread of sunspots from the poles of the Sun to the equator of the Sun that form a "butterfly" pattern.

Carbon Cycle: The global scale-exchange of carbon among the Earth's reservoirs of carbon storage, namely the atmosphere, oceans, vegetation, soils, and geologic deposits and minerals. This involves components in food chains, in the atmosphere as carbon dioxide, in the hydrosphere and the geosphere.

Carbon Dioxide (CO^2): The greenhouse gas whose concentration is being most affected directly by human activities. Carbon dioxide also serves as the reference to compare all other greenhouse gases. One of the sources of carbon dioxide emissions is fossil fuel combustion. Carbon dioxide emissions are also a product of forest clearing, biomass burning, and non-energy production processes such as cement production.

Carbon Dioxide Equivalent (CDE): A metric measure used to compare the emissions from various greenhouse gases based upon their global warming potential (GWP). Carbon dioxide equivalents are commonly expressed as "million metric tons of carbon dioxide equivalents" (MMTCDE) or "million short tons of carbon dioxide equivalents" (MSTCDE). The carbon dioxide equivalent for a gas is derived by multiplying the tons of the gas by the associated GWP. MMTCDE = (million metric tons of a gas) x (GWP of the gas). For example, the GWP for methane is 24.5. The means that emissions for one million metric ton of methane is equivalent to emissions of 24.5 million metric tons of carbon dioxide. Carbon may also be as the reference to covert other greenhouse gases to carbon equivalents. To convert carbon to carbon dioxide, multiply the carbon by 44 ½ (the ratio of the molecular weight of carbon to carbon dioxide).

Carbon Dioxide Fertilization: An expression (sometimes reduced to "carbon fertilization") used to denote increased plant growth due to a higher carbon dioxide concentration in the atmosphere.

Carbon Equivalent (CE): A metric measure used to compare the emissions of the different greenhouse gases based upon their global warming potential (GWP). Greenhouse gas emissions in the U.S. are most commonly expressed "a million metric tons of carbon equivalents"

(MMTCE). Global warming potentials are used to convert greenhouse gases to carbon dioxide equivalents. Carbon dioxide equivalents can then be converted to carbon equivalents by multiplying the carbon dioxide equivalents by 12/44 (the ratio of the molecular weight of carbon to carbon dioxide). Thus the formula to derive carbon equivalent is: MMTCE = (million metric tons of gas) x (GWP of the gas) x (12/44).

Carbon Sequestration: The uptake and storage of carbon. Trees and plants, for example, absorb carbon dioxide, release the oxygen, and store the carbon. Fossil fuels were at one time biomass and continue to store the carbon until burned.

Carbon Sinks: Carbon reservoirs and conditions that take in and store more carbon (carbon sequestration) than they release. Carbon sinks can serve to partially offset greenhouse gas emissions. Forests, soils, and oceans are common carbon sinks.

Carbon Years: A method of dating based upon the radioactive decay of certain types of carbon atoms. Also referred to as Radio Carbon Dating.

Celsius (C): Named after the inventor Anders Celsius (1701-1744), a Swiss Astronomer. Specifically, a designation for a scale of measurement or a thermometer where 0° is the freezing point of water and 100° is the boiling point of water. The formula for converting degrees Celsius to degrees Fahrenheit is: F° = 9/5C° +32.

CH^4: The chemical symbol for methane.

Chicxulub Crater: An undersea asteroid crater located off the Yucatan Peninsula. It is 200 miles in diameter, and it is believed to be the crater of the asteroid that caused the extinction of the dinosaurs.

Chlorofluorocarbons (CFCs) and Related Compounds: This family of anthropogenic compounds includes chlorofluorocarbons (CFCs), bromofluorocarbons (halons), methyl chloroform, carbon tetrachloride, methyl bromide, and hydrochlorofluorcarbons (HFCs). These compounds have been shown to deplete stratospheric ozone, and therefore are typically referred to as ozone depleting substances.

Climate: The average weather (usually taken over a 30-year period of time) for a particular region and time period. Climate is not the same as weather, but rather, it is the average long-term pattern of weather for a particular region. Weather describes the short-term state of the atmosphere. Climatic elements include precipitation, temperature, humidity, sunshine, wind velocity, phenomena such as fog, frost, hail storms, and other measures of the weather.

Climate Change (also referred to as "global climate change"): A departure from the expected average weather or climate norms. The term "climate change" is sometimes used to refer to all forms of climatic inconsistency, but because the Earth's climate is never static, the term is more properly used to imply a significant change from one climatic condition to another. In some cases, "climate change" has been used synonymously with the germ global warming. Scientists, however, tend to use the term in the wider sense to also include natural changes in the climate.

Climate Change Action Plan (CCAP): Unveiled in October 1993 by President Clinton, the CCAP is the U.S. plan for meeting its pledge to reduce greenhouse gas emissions under the terms of the Framework Convention on Climate Change (FCCC). The goal of the CCAP is to reduce U.S. emissions of anthropogenic greenhouse gases to 1990 levels by the year 2000.

Climate Feedback: An atmospheric, oceanic, terrestrial, or other process that is activated by the direct climate change induced by changes in radiative forcing. Climate feedbacks may increase (positive feedback) or diminish (negative feedback) the magnitude of the direct climate changes.

Climate Forcing: Natural phenomena that cause (or "force") a change in the Earth's climate.

Climate Lag: The delay that occurs in climate change as a result of some factor that changes only very slowly.

Climate Model: A quantitative way of representing the interactions of the atmosphere, oceans, land surface, and ice. Models can range from relatively simple to quite comprehensive.

Climate Modeling: The simulation of the climate using computer-based models.

Climate Sensitivity: The equilibrium response of the climate to a change in radiative forcing; for example, a doubling of the carbon dioxide concentration.

Climate System (or Earth System): The atmosphere, the oceans, the biosphere, the cryosphere, and the geosphere, together make up the climate system.

Cloud Condensation Nuclei (CNN): Generally referring to certain types of sulfur-based aerosols that tend to create clouds.

CO^2: The chemical symbol for carbon dioxide.

CNN: *See*, Cloud Condensation Nuclei

Conductivity-Temperature-Depth Recorder (CTD): An instrument used by physical oceanographers to measure the salinity and temperature of the ocean at various depths.

Cogeneration: The process by which two different and useful forms of energy are produced at the same time. For example, while boiling water to generate electricity, the leftover steam can be sold for industrial processes or space heating.

Compost: Decayed organic matter that can be used as a fertilizer or soil additive.

Conference of the Parties (CoP): The CoP is the collection of nations which have ratified the Framework convention on Climate Change (FCCC), currently over 150 strong, and about 50 Observer States. The primary role of the CoP is to keep the implementation of the Convention under review and to take the decisions necessary for the effective implementation of the Convention. The first CoP (CoP1) took place in Berlin from March 28th to April 17th, 1995.

Corona: Outer layer of the Sun's atmosphere.

Cryosphere: The frozen part of the Earth's surface. The cryosphere includes the polar ice caps, continental ice sheets, mountain glaciers, sea ice, snow cover, lake and river ice, and permafrost.

Dansgaard-Oeschger Events: Warming events identified from the Greenland ice core data that occurred generally during the last ice age.

Deforestation: Those practices or processes that result in the change of forested lands to non-forest uses. This is often cited as one of the major causes of the enhanced greenhouse effect for two reasons: 1) the burning or decomposition of the wood releases carbon dioxide; and 2) the trees that once removed carbon dioxide from the atmosphere in the process of photosynthesis are no longer present and contributing to carbon storage.

Dentachronology: The study of tree rings to determine past events.

Desertification: The progressive destruction or degradation of existing vegetative covers to form deserts. This can occur due to overgrazing, deforestation, drought and the burning of extensive area of land due to slash-and-burn farming, grassland and forest fires, and the like. Once formed, deserts can only support a sparse range of vegetation. Climatic effects associated with this phenomenon include increased albedo, reduced atmospheric humidity, and greater atmospheric dust (aerosol) loading.

Drag (as used in reference to orbital bodies): The frictional interaction with the orbiting body and the atmosphere of a planet (usually that of Earth) or other astronomic body.

Eccentricity in Earth's Orbit around the Sun: The amount that the Earth's orbit around the Sun varies from a perfect circle to that of an ellipse, an amount that varies over time from 1% to 5% in orbital variation. Over a 95,000-year cycle, the Earth's orbit around the Sun changes from a thin ellipse (oval) to a circle and back again. When the orbit around the Sun is most elliptical, there is as much as a 30% difference in the distance between the Earth and the Sun at perihelion and aphelion. Though the current three million mile difference in distance doesn't change the amount of solar energy we receive much, a 30% difference really would modify the amount of solar energy received and would make perihelion much warmer time of the year than aphelion.

Eemian: The last interglacial period, or warm period between ice ages, before the present interglacial. Is believed to have occurred between 110,000 and 120,000 y.b.p. It is named after a terrestrial pollen record of the Netherlands in a geological known as the Sangamon Stage on land in the U.S. It was followed by the colder last glacial period known as the last Ice Age, followed by another interglacial called the Holocene that began approximately 11,000 y.b.p. and continues through to the present.

Electrical Conductivity Measurement (ECM): An analytical process to determine the chemical composition of an ice core.

El Niño/Southern Oscillation (ENSO): A climatic phenomenon occurring irregularly, but generally every three to five years. El Niños often first become evident during the Christmas season (El Niño means Christ child) in the surface oceans of the eastern tropical Pacific Ocean. The phenomenon involves seasonal changes in the direction of the tropical winds over the Pacific and abnormally warm surface ocean temperatures. The changes in the tropics are most intense in the Pacific Region. These changes can disrupt weather patterns throughout the tropics and can extend to higher latitudes, especially in Central and North America. The relationship between these events and global weather patterns are currently the subject of much research in order to enhance prediction of seasonal to interannual fluctuations in the climate. El Niños are an area of the Pacific Ocean one and a half times the size of the continental U.S. that becomes warmer than normal by as much a 10° F. This phenomenon last occurred in 1997-98, when it was exceptionally strong and disrupted climate around the world. La Niña is the cold water counterpart that often follows an El Niño.

Emissions: The release of a substance (usually a gas when referring to the subject of climate change) into the atmosphere.

Enhanced Greenhouse Effect: Manmade contributions to the greenhouse effect.

Eolian: Literally "wind blown." A term used in geology to describe deposits made by the wind, such as sand dunes.

Equinox: The time of year that the Sun is positioned directly over the equator, currently occurring at approximately March 21 and September 21.

Evapotranspiration: The sum of evaporation and plant transpiration. Potential evapotranspiration is the amount of water that could be evaporated, or transpired, at a given temperature and humidity, if there was plenty of water available. Actual evapotranspiration cannot be any greater than precipitation, and will usually be less because some water will run off in rivers and flow to the oceans. If potential evapotranspiration is greater than actual precipitation, then soils are extremely dry during at least a major part of the year.

Fahrenheit (F): From its inventor, Daniel Fahrenheit (1686-1736), a German physicist. Specifically, a designation for a scale of measurement or a thermometer where 32° is the freezing point of water and 212° is the boiling point of water. The formula for converting degrees Fahrenheit to degrees Celsius is: $C° = 5/9 (F°-32)$. *See also*, Celsius and Kelvin Scale.

Feedbacks: Positive feedbacks work to accelerate or amplify, and negative feedbacks work to slow down or offset an event or occurrence such as global warming.

Feedback Mechanisms: A mechanism that connects one aspect of a system to another. The connection can be either amplifying (positive feedback) or moderating (negative feedback).

Fertilization: A term used to denote efforts to enhance plant growth by increased application of nitrogen-based fertilizer or increased deposition of nitrates in precipitation.

Fingerprints: Indicators of global, long-term warming trend observed in this historical record. They include heat waves, sea level rise, melting glaciers and warming of the poles.

Fluorocarbons: Carbon-fluorine compounds that often contain other elements such as hydrogen, chlorine, or bromine. Common fluorocarbons include chlorofluorocarbons and related compounds (also known as ozone depleting substances), hydrofluorocarbons (HFCs) and perfluorocarbons (PFCs).

Forcing Mechanism: A process that alters the energy balance of the climate system, i.e., changes the relative balance between incoming solar radiation and outgoing infrared radiation from Earth. Such mechanisms include changes in solar irradiance, volcanic eruptions, and enhancement of the natural greenhouse effect by emission of carbon dioxide.

Fossil Fuel: A general term for combustible geologic deposits of carbon in reduced (organic) form and of biological origin, including coal, oil, natural gas, oil shale, and tar sands.

Fossil Fuel Combustion: Burning of coal, oil (including gasoline), or natural gas. This burning, usually to generate energy, releases carbon dioxide, as well as combustion by-products that can include unburned hydrocarbons, methane and carbon monoxide. Carbon monoxide, methane, and many of the unburned hydrocarbons slowly oxidize into carbon dioxide in the atmosphere. Common sources of fossil fuel combustion include automobiles and electric utilities.

Framework Convention on Climate Change (FCCC): An international treaty unveiled at the United Nations Conference on Environment and Development (UNCED) also known as the "Rio Summit"), in June 1992. The FCCC commits signatory countries to stabilize anthropogenic greenhouse gas emissions to levels that would prevent alleged anthropogenic interference with the climate system.

GCR: Galactic cosmic ray flux.

General Circulation Model (GCM): Computer programs designed to study and predict climate. Using the basic laws of science (conservation of mass and momentum, etc.), these programs represent a global, three-dimensional computer model of the climate system, which can be used to simulate human-induced climate change. GCMs are highly complex and they represent the effects of such factors as reflective and absorptive properties of atmospheric water vapor, greenhouse gas concentrations, clouds, annual and daily solar heating, ocean temperatures and ice boundaries. The most recent GCMs include global representations of the atmosphere, oceans, and land surface.

Geosphere: The soils, sediments, and rock layers of the Earth's crust, both continental and beneath the ocean floors.

GIGO: Garbage in equals garbage out. Computer term that means the results from a computer program is only as good as the inputted data. Bad data equals bad results.

GHCN: Global Historical Climatology Network, created by the

National Climate Data Center of NOAA.

Gigatons: One billion tons.

Gleissber Period: A 70 to 80 year oscillation of the 11-year solar sunspot cycle. That is, while sunspots appear on an approximate 11-year cycle, the intensity (or amplitude) of those 11-year cycles also goes through a 70 to 90 year cycle. A solar sunspot cycle that occurs at the height of a Gleissber Period is significantly more intense than a solar sunspot cycle that occurs at the bottom of the Gleissber Period.

Global Warming: Refers to the observation that the atmosphere near the Earth's surface temperature on a global, or worldwide scale, is warming without any implications for the cause or magnitude. Global warming (as well as global cooling) has occurred in the distant past as the result of natural influences, but the term is most often used to refer to the warming predicted to occur as a result of increased emissions of greenhouse gases.

Global Warming Potential (GWP): The index used to translate the level of emissions of various gases into a common measure in order to compare the relative radiative forcing of different gases without directly calculating the changes in atmospheric concentrations. GWPs are calculated as the ratio of the radiative forcing that would result from the emissions of one kilogram of a greenhouse gas to that from emission of one kilogram of carbon dioxide over a period of time (usually 100 years).

Greenhouse Effect: The effect produced as greenhouse gases allow incoming solar radiation to pass through the atmosphere, but prevent most of the outgoing infrared radiation from the surface and lower atmosphere from escaping into outer space. This process occurs naturally and has kept the Earth's temperature about 59° F warmer than it would otherwise be. Current life on Earth could not be sustained without the natural greenhouse effect. It is also a term that describes how water vapor, carbon dioxide, and other gases in the atmosphere help maintain the temperature at the Earth's surface.

Greenhouse Gas: Any gas that absorbs infrared radiation in the atmosphere. Greenhouse gases include water vapor (H^2O), carbon dioxide (CO^2), methane (CH^4), nitrous oxide (N^2O), halogenated fluoluoro-carbrocarbons (HCFCs), ozone (O^3), perfluorinatedcarbons (PFCs), and hydrofluorocarbons (HFCs).

Greenhouse Warming: The warming of the Earth as concentrations of greenhouse gases in the atmosphere increase much like what happens when the windows of a greenhouse (or an automobile) are closed on a warm, sunny day.

Greenland Ice Core Project (GRIP): An ice core drill site located 30 km to the east of the Greenland Ice Sheet Project Two that penetrated the ice sheet to a depth of 3,028.8 meters.

Greenland Ice Sheet Project Two (GISP2): An ice core drill site based on the Greenland Ice Sheet in the Summit Region of Central Greenland. GISP2 recovered the deepest ice core record in the world for the Northern Hemisphere (3053.44 meters).

Gyre: From the Greek word, gyros and meaning circle. A circular wind pattern. In oceanography, it refers to the giant circular currents of Earth's oceans. Those circular currents flow clockwise in the Northern Hemisphere and they flow counterclockwise in the Southern Hemisphere.

H^2O: The chemical symbol for water.

Halocarbons: Chemicals consisting of carbon, sometimes hydrogen, and either chlorine, fluorine bromine, or iodine.

Halons: These manmade substances (also known as bromofluorocarbons) are chlorofluorocarbons that contain bromine.

Harbingers: Events that foreshadow the impacts likely to become more frequent and widespread with continued global warming. They include, spreading disease, earlier spring arrival, plant and animal range shifts, coral reef bleaching, shorter periods of rain with greater intensity, and droughts and fires.

Heat Island: A geographic area that causes artificially high temperature readings due to manmade structures or unusual geographic features.

Heinrich Event: A sudden and severe drop in global temperature of 3° to 6° C during the already glacial climate of an ice age, originating in the North Atlantic.

Heliosphere: Having to do, and relating to, the area of outer space influenced by the Sun.

Herbivore: An animal that eats only plants as compared to a carnivore, an animal that eats only meat, or an omnivore, an animal that eats both plants and meat.

Higher Latitude: An expression used to indicate an object, geographic location, or climatic event is closer to either the North or South Poles than it is to the Equator, as opposed to the expression "lower latitude" meaning closer to the Equator than to one or the other of the poles.

Holocene: The current geological period in which we all live, a period that began approximately 11,500 y.b.p. It is also defined geologically as an interglacial period.

Holocene Optimum: A period between 9,000 and 5,000 y.b.p., when the global climate conditioners were warmer and more moist than today.

Hydrological Cycle: The cycle of evaporation from the Earth and condensation of water in clouds that creates rainfall.

Hydrosphere: The part of the Earth composed of water including clouds, oceans, seas, ice caps, glaciers, lakes, rivers, underground water supplies, and atmospheric water vapor.

Hydroxyls: Water molecules contain only one atom of hydrogen and one atom of oxygen (HO), whereas water normally contains two atoms of hydrogen and one atom of oxygen (H^2O). Because hydroxyls lack a hydrogen atom, they are unstable and seek another atom for chemical balance. As such, it is believed hydroxyls in the atmosphere play a significant role in cleansing the atmosphere of potential greenhouse gases.

Ice Ages: Intervals of time when large areas of the surface of the Earth are covered with ice sheets (large continental glaciers). In addition, the term "Ice Age" is sometimes used to refer to the last major glaciations that occurred in North America and Eurasia. When used in this way, the first letters of both words are capitalized.

Ice Core: A cylindrical section of ice removed from a glacier or an ice sheet in order to study climate patterns of the past. By performing chemical analyses on the air trapped in the ice, scientists can estimate the percentage of carbon dioxide and other trace gases in the atmosphere in the past.

Infrared Radiation: Heat energy that is emitted from all solids, liquids, and gases. In the context of the greenhouse issue, the term refers to the heat energy emitted by the Earth's surface and its atmosphere. Greenhouse gases strongly absorb this radiation in the atmosphere, and reradiate some back towards the surface as infrared radiation, creating the greenhouse effect.

Interglacial: A period of warm global climate occurring between ice ages. We are currently in an interglacial period called the Holocene that began about 11,500 y.b.p. The previous interglacial before the one we are in was called the Eemian. *See also*, Interstadials.

Intergovernmental Panel on Climate Change (IPCC): The IPCC was established jointly by the United Nations Environment Programme and the World Meteorological Organization in 1988. The purpose of the

IPCC is to assess information in the scientific and technical literature related to all significant components on the issue of climate change.

Interstadials: Sudden and short-lived warm conditions that occur during the generally colder global climate conditions of an ice age, as compared to those colder conditions, which are know as stadials.

Ionosphere: The layer of air that extends from 50 to 300 miles above the surface of the Earth.

IR: Infrared. *See* Infrared Radiation.

Joint Implementation: Agreements made between two or more nations under the auspices of the Framework Convention on Climate Change (FCCC) to help reduce greenhouse gas emissions.

K.a.: Kilo years ago. The measurement of years in increments of 1,000 years. 10 k.a.=10,000 years.

Kelvin Scale (K): From Baron Kelvin (1824-1907), its inventor, British Physicist and Mathematician. A scale that measures temperature in degrees Celsius with Absolute Zero (-273.15° C) as its starting point. For example, the freezing point of water which is 0° C (32° F) is 273.15° K, and the boiling point of water which is 100° C (212° F) is 373.15° K. The Kelvin Scale is generally used to measure very large temperatures such as the temperature of the Sun. *See also*, Fahrenheit and Celsius.

Krakatoa: An Indonesian volcano that erupted in 1883.

Last Glacial Maximum: The point at which the global expanse of ice was the greatest during the last ice age, about 21,000 y.b.p. It is also referred to as the LGM.

Lifetime (Atmospheric): The lifetime of a greenhouse gas refers to the approximate amount of time it would take for the anthropogenic increment to an atmospheric pollutant concentration to return to its natural level (assuming emissions cease) as a result of either being converted to another chemical compound or being taken out of the atmosphere by a sink. This time depends on the pollutant's sources and sinks as well as it reactivity. The lifetime of a pollutant is often considered in conjunction with the mixing of pollutants in the atmosphere; a long lifetime will allow the pollutant to mix throughout the atmosphere. Average lifetimes can vary from about a week (sulfate aerosols) to more than a century (CFCs, and carbon dioxide).

Little Ice Age (also called the Mini Ice Age): A period of unusual cooling that occurred in Europe and North America (and possibly elsewhere globally) between approximately the years of 1350 and 1900 (although the exact dates are in dispute).

Lower Pleniglacial: A point of extreme cold conditions of the ice age that occurred around 70,000 y.b.p.

Mauna Loa: A volcano on the Island of Hawaii in the State of Hawaii where scientists have maintained the longest continuous collection of reliable daily atmospheric records.

Mean Sea Level: The average height of the Earth's oceans.

Medieval Warm Period: An unusually warm period that occurred in Europe between the years 1000 and 1300.

Meta Analysis: A study and analysis of a large volume of publications on a particular subject (usually involving complex issues and opinions) in an effort to determine if there is a consensus, or trend within the publications on that subject.

Meteorology: The science of weather and climate-related phenomena.

Methane (CH^4): A hydrocarbon that is a greenhouse gas with a global warming potential most recently estimated at 24.5. Methane is produced through anaerobic (without oxygen) decomposition of waste landfills, animal digestion, decomposition of animal wastes, production and distribution of natural gas and oil, coal production, and incomplete fossil fuel combustion.

Metric Ton: Common international measurement for the quantity of greenhouse gas emissions. A metric ton is equal to 2,205 lbs. or 1.1 short tons.

Milankovitch, Milan: Developed a theory of solar variation as the cause of climate variation on Earth.

MTCDE: Million Metric Tons of Carbon Equivalents.

Mount Pinatubo: A volcano in the Philippine Islands that erupted in 1991. The eruption of Mt. Pinatubo ejected enough particulate and sulfate aerosol matter into the atmosphere to block some of the incoming solar radiation from reaching the atmosphere. This effectively cooled the planet from 1992 to 1994.

MSL: *See* Mean Sea Level.

MSTCDE: Million Short Tons of Carbon Dioxide Equivalents.

N^2O: The chemical symbol for nitrous oxide.

NASA: National Aeronautics and Space Administration.

Nitrogen Oxides (Nox): Gases consisting of one molecule of nitrogen and varying numbers of oxygen molecules. Nitrogen oxides are produced in the emissions of vehicle exhausts and from power stations. In the atmosphere, nitrogen oxides can contribute to formation of photochemical ozone (smog), can impair visibility, and have health consequences; they are thus considered pollutants.

Nitrous Oxide (N^2O): A powerful greenhouse gas with a global warming potential of 320. Major sources of nitrous oxide include soil cultivation practices, especially the use of commercial and organic fertilizers, fossil fuel combustion, nitric acid production, and biomass burning.

NOAA: National Oceanic and Atmospheric Administration.

North Atlantic Oscillation: A seesawing variance in the atmospheric pressure over the Arctic and the subtropical North Atlantic Ocean. It determines sea surface temperatures and winter weather in the area. When the pressure is lower than normal over the Arctic, the eastern U.S. and northwestern Europe experience mild winters. In its opposite phase, winters are mild in eastern Canada and Greenland.

O^3: Chemical symbol for ozone.

Obliquity: A 42,000 year cycle where the Earth wobbles and the angle of its axis, with respect to the plane of revolution around the Sun, varies between 22.1° and 24.5°. Less of an angle than our current 23.45 ° means less seasonal differences between the Northern and Southern Hemispheres while a greater angle means greater seasonal differences (i.e., a warmer summer and cooler winter).

Ockham's Razor: A principle that says the simple explanation is the best.

Ozone (O^3): Ozone consists of three atoms of oxygen bonded together in contrast to normal atmospheric oxygen, which consists of two atoms of oxygen. Ozone is an important greenhouse gas found in both the stratosphere (about 90% of the total atmospheric loading) and the troposphere (about 10%). Ozone has other effects beyond acting as a greenhouse gas. In the stratosphere, ozone provides a protective layer shielding the Earth from ultraviolet radiation and subsequent harmful health effect on humans and the environment. In the troposphere, the oxygen molecules in ozone combine with other chemicals and gases (oxidization) to cause smog.

Ozone Depletion Potential (ODP): A rating system used to determine the destructive capability of CFC's on the ozone layer in the atmosphere. In technical terms, ODP is equal to the cumulative ozone depletion of the compound (causing the depletion) divided by the ozone depletion caused by the release of an equal amount of CFC-12, one of the first CFC's.

Ozone Holes: Areas in the stratosphere where the ozone has been destroyed by various chemical reactions.

Pacific Decadal Oscillation: A long-term oscillation (lasting decades) in the ocean temperatures of the North Pacific Ocean. This long term change in the sea surface temperatures seems to have two phases. From about 1977 to 1997, temperatures in the middle and western part of the North Pacific Ocean were cooler that average, while waters off the western U.S. were warmer. At the end of the 1990s, however, temperatures flip-flopped perhaps signaling a return to the conditions that were last prevalent between 1947 and 1976.

Palmer Drought Severity Index (PDSI): One of several scales used to measure the severity of a drought.

Parsimony: A principle of scientific investigations that states when confronted with two, or more, complex solutions to a problem, the most simple (least complex) is the solution that will prevail.

Particulates: Tiny pieces of solid or liquid matter in the atmosphere, such as soot, dust, fumes, or mist.

Perferfluorocarbons (PFCs): A group of manmade chemicals composed of carbon and fluorine only: CF^4 and $C2F^6$. These chemicals, specifically CF^4 and $C2F^6$ (along with hydrofluorocarbons) were introduced as alternatives to the ozone depleting substances. In addition, they are emitted as by-products of industrial processes and are also used in manufacturing. PFCs do not harm the stratospheric ozone layer, but they are powerful greenhouse gases: CF^4 has a global warming potential of 6,300 and $C2F^6$ has a GWP of 12,500.

Paleoclimate: Climate that existed before humans began collecting instrumental measurements of weather.

Paleoclimotologist: A person who studies paleoclimatology, climate before recorded history.

Paleoclimatology: The study of past climate before recorded records and the investigation of the climate processes underlying those conditions.

Paradigm: World view. Specifically, the way a particular a scientific discipline perceives the world and, generally, everything else.

Parasol Effect: The effect in blocking sunlight that aerosols released by volcanic eruptions has. Generally, those aerosols block the sunlight and can have a cooling effect on global temperatures for several years, the length of which is dependent upon the latitude at which the volcano is located.

Pathogen: Anything that causes or transmits diseases into plants, animals, and human beings.

Perihelion: The point the Earth's orbit (or the orbit of any other stellar body) is closest to the Sun (or the object around which it orbits).

Photodissociation: The process by which a chemical combination breaks up into simple constituents due to exposure to the Sun's radiation.

Phenomena Glacial Rebound (PGR): The effect of land once covered by ice age glaciers rising (or rebounding) when the weight of those glaciers is removed by melting or withdrawal.

Photosynthesis: The process by which green plants use light to synthesize organic compounds from carbon dioxide and water. In the process, oxygen and water are released. Increased levels of carbon dioxide can increase net photosynthesis in some plants. Plants create a very important reservoir for carbon dioxide.

Plate Tectonics: The theory that the Earth's surface consists of plates that move and accountant for continental drift, mountains and other geological features.

Pollutant: Strictly, too much of any substance in the wrong place or at the wrong time is a pollutant. More generally, an atmospheric pollution may be defined as the presence of substances in the atmosphere, resulting from manmade activities or from natural processes that cause adverse effects to human health, property, and the environment.

Polychaetes: Marine worms.

Ppb: Parts per billion.

Ppm: Parts per million.

Precautionary Approach: The approach promoted under the Framework Convention of Climate Change to help achieve stabilization of greenhouse gas concentrations in the atmosphere at a level that would prevent dangerous interference with the climate system.

Precession: Changes in the timing as to when solar equinoxes occur. There is a tendency of the Earth's axis to wobble in space over a period of 23,000 years. The Earth's precession is one of the factors that results in the planet receiving different amounts of solar energy over extended periods of time.

Proxy Records (also called Proxy Climate Data): The use of natural environmental records to infer past climate conditions. Examples of such proxy records in the study of paleoclimatology are ice cores, tree rings, ocean and lake sediments, ocean reefs, and geological formations.

Radiation: Energy emitted in the form of electromagnetic waves. Radiation has different characteristics depending upon the wavelength. Because the radiation from the Sun is relatively energetic, it has a short wavelength (ultraviolet, visible light, and near infrared) while energy reradiated from the Earth's surface and the atmosphere has a longer wavelength (infrared radiation) because the Earth is cooler than the Sun.

Radiative Forcing: A change in the balance between incoming solar radiation and outgoing infrared radiation. Without any radiative forcing, solar radiation coming to the Earth would continue to be approximately equal to the infrared radiation emitted from the Earth. The addition of greenhouse gases traps an increased fraction of the infrared radiation, re-radiating it back toward the surface and creating a warming influence (i.e., positive radiative forcing because incoming solar radiation will exceed outgoing infrared radiation).

Residence Time: The average time spent in a reservoir (such as a reservoir of gas like the atmosphere) by an individual atom or molecule. Also, the age of a molecule when it leaves the reservoir. With respect to greenhouse gases, residence time usually refers to how long a particular molecule remains in the atmosphere.

Resolution: The accuracy as measured in years to which an event can be said to have occurred. High-resolution means a determination can be made within a few years and low resolution means a determination can be made only within wide latitude of years.

Respiration: The process by which animals use up stored foods (by combustion with oxygen) to produce energy.

SAM: Solar Activity Magnifier.

Scenario: A plan, calculation, computer model, and the like, designed to suggest what future events might be. It is not a predication. Rather it is a way of investigating the implications of particular assumptions about future trends.

SCICEX: Science Ice Expeditions to the Arctic Ocean.

SEGMs: Solar Enhanced Greenhouse Mechanisms.

Shocked Quartz: Mineral quartz with stress lines indicating it was subjected to a high force of energy (such as an explosion) at sometime in the past.

Short Ton: Common measurement for a ton in the United States. A short ton is equal to 2,000 lbs., or 0.907 metric tons.

Sink: Any biological or geological feature that absorbs (removes) carbon dioxide from the atmosphere. A reservoir that uptakes a pollutant from another part of its cycle. Soil and trees tend to act as natural sinks for carbon. Also, any process that absorbs or destroys greenhouse gases.

Snowball Earth Theory: A theory that states the Earth completely froze over and then thawed several times between 750 million and 500 million years ago.

Solar Extreme Ultraviolet (EUV): The increased output of ultraviolet radiation that occurs during a solar maximum.

Solar Maximum: The maximum of sunspot activity that occurs on an approximate 11-year cycle.

Solar Radiation: Energy from the Sun. Also referred to as short wave radiation. Of importance to the climate system, solar radiation includes ultraviolet radiation, visible light radiation, and infrared radiation.

Solar Sunspot Cycle: The approximate 11-year cycle in the minimum and maximum occurrence of sunspots on the Sun.

Stadials: The generally colder global climate conditions of an ice age.

Stratosphere: The part of the atmosphere directly above the troposphere. The layer of air that extends from the upper level of the Troposphere to 50 miles above the surface of the Earth.

Stomates: The small openings in leaf surfaces of plants through which carbon dioxide is absorbed and water vapor is released.

Stock Air Pollutant: An air pollutant that has a long lifetime in the atmosphere, and therefore can accumulate over time. It is also generally well mixed in the atmosphere.

Sulfate Aerosol: Particulate matter that consists of compounds of sulfur formed by the interaction of sulfur dioxide and sulfur trioxide with other compounds in the atmosphere. Sulfate aerosols are injected into the atmosphere by the combustion of fossil fuels and from the eruption of volcanoes like Mt. Pinatubo. Recently theory suggests that sulfate aerosols may lower the Earth's temperatures by reflecting away solar radiation (negative radiative forcing).

Sulfur Dioxide (SO^2): A compound composed of one sulfur and two oxygen molecules. Sulfur dioxide emitted into the atmosphere through natural and anthropogenic processes. It is changed in a complex series of chemical reactions to sulfate aerosols. These aerosols result in a negative radiative forcing (i.e., tending to cool the Earth's surface).

Sulfur Hexafluoride (SF^6): A very powerful greenhouse gas used primarily in electrical transmission and distribution systems. Sulfur hexafluoride has a global warming potential of 24,900.

Tambora: A volcano in Indonesia that erupted in 1815. It is believed this was the largest volcanic eruption over the last 5,000 years. The year following the eruption has been called the "Year without a Summer." In June and July of 1816, New England and northern Europe suffered frost and snow as a result of the material ejected into the atmosphere by this volcanic eruption.

Teleconnections: As used in relation to the El Niño/Southern Oscillator, are the corresponding effects the El Niño/Southern Oscillator have on climate.

Teragrams: One teragram equals one metric ton.

Thermohaline Circulation: The name given to ocean currents that are driven by differences in the heat and salt content of seawater.

Tilt of Earth's Axis: The amount in degrees that the Earth's North Pole "tilts" towards or away from the Sun.

Toba: An Indonesian volcano that erupted approximately 70,000 y.b.p. The ejection of material from this volcano was so great that it blocked the Sun from the Earth for more than 10 years, and that cooled Earth's global temperature to near winter conditions. It is estimated that the worldwide human population of that time, some 1,000,000 individuals, was reduced to a population of fewer than 5,000 people.

Total Ozone Mapping Spectrometer (TOMS): A satellite borne instrument used to gain global measurements of ozone levels.

Trace Gas: Any one of the less common gases found in the atmosphere. Nitrogen, oxygen, and argon make up more than 99% of the atmosphere. Other gases such as carbon dioxide, water vapor, methane, oxides of nitrogen, ozone, and ammonia, are considered trace gases.

Tropical Atlantic Variability: Variations in sea surface temperatures in the Atlantic Ocean that affect precipitation patterns in parts of South America and Africa. When water north of the Equator is colder than water to the south, northeast Brazil receives higher than average rainfall, and the Sahel experiences drought. Reversals in sea surface temperatures bring opposite precipitation patterns.

Troposphere: The lowest layer of the atmosphere. The troposphere extends from the Earth's surface up to about 10-15 km. It is defined as the layer of air that extends from the surface of the Earth to the area where the air's temperature stops becoming lower. This height is approximately 5 miles high over the North and South Poles, and it is approximately 10 miles high over the Equator.

Tropospheric Ozone (O^3): Ozone that is located in the troposphere and plays a significant role in the greenhouse gas effect and urban smog. *See also*, Ozone.

Tropospheric Ozone Precursor: Gases that influence the rate at which ozone is created and destroyed in the atmosphere. Such gases include carbon monoxide (CO), nitrogen oxides (Nox), and non-methane volatile organic compounds (NMVOCs).

UCS: The Union of Concerned Scientists.

UNEP: United Nations Environment Programme.

Upper Pleniglacial: A point of extreme cold conditions of the last Ice Age that occurred sometime between 21,000-17,000 yba, also called the Late Glacial Cold State.

WAIS: West Antarctic Ice Sheet.

Water Vapor: The most abundant greenhouse gas, it is the water present in the atmosphere in gaseous form. Water vapor is an important part of the natural greenhouse effect. While mankind is not significantly increasing its concentration, it contributes to the enhanced greenhouse effect because the warming influence of greenhouse gases leads to a positive water vapor feedback. In addition to its role as a natural greenhouse gas, water vapor plays an important role in regulating the temperature of the planet because clouds form when excess water vapor in the atmosphere condenses to form ice, water droplets, and precipitation.

Weather: The state of the atmospheric conditions (i.e.), hot/cold, wet/dry, calm/stormy, sunny/cloudy) that exist over relatively short periods of time (hours to a couple of days) within a specific geographic area. Weather is the specific condition of the atmosphere at a particular place and time. It is measured in terms of such things as wind, temperature, humidity, atmospheric pressure, cloudiness, and precipitation. In most places, weather can change from hour-to-hour, day-to-day, and season-to-season. Climate is the average long-term of weather over time and space.

Wisconsin: The name for the ice sheet that covered North America during the last Ice Age.

Würm: The name for the ice sheet that covered Europe during the last Ice Age.

Younger-Dryas Event (YD): A period of sudden, rapid cooling of global climate temperature, occurring between 12,900 to 11,500 y.b.p. resulting in widespread cool, dry conditions lasting approximately 400 years.

Ziryanskaya: The name for the ice sheet that covered Siberia during the last Ice Age.

Bibliography

Abadie, Victor H. (3 Oct. 1997.) "We're Between Ice Ages, So Relax." Wall Street Journal, A11. http://www.ssc.msu.edu/~geo/stu/duda/gw.html (9/4/00).

Adams, Jonathan. (1999). "An Inventory of Data for Reconstructing 'Natural Steady State' Carbon Storage in Terrestrial Ecosystems." Environment Sciences Division, Oak Ridge National Laboratory, Oak Ridge, Tennessee. http://www.esd.ornl.gov/projects/qen/nerc.html (9/3/00).

Adams, Jonathan, Mark Maslin, and Ellen T. Thomas. (1999). "Sudden Climate Transitions During the Quaternary." Progress in Physical Geography. http://www.esd.ornl.gov/projects/(9/3/00).

Ackerman, Jennifer. (Oct. 2, 2000). "New Eyes on the Ocean" National Geographic, Vol. 198, No. 4, pp. 86-115.

Angell, James K. (2000). "Stratospheric Warming Following Volcanic Eruptions." Air Resources Laboratory/NOAA, Silver Spring, Maryland. http://capita.wustl.edu/ (10/14/00).

Ball, Philip. (2 Dec. 1999). "Climate: An Ocean Switch for Global Cooling." Nature News Service, Macmillan Publishers, Ltd., England. http://helix.nature.com/ 8/31/00).

Balling, Jr., Robert C. (2000). "Not in Kansas Anymore?" Arizona State University. http://www.Greeningearthsociety.org/climate/ 9/12/00).

Barber, Don. (21 July 1999). "Catastrophic Draining of Huge Lakes Tied to Ancient Global Cooling Event" University of Colorado, Boulder, Colorado. http://www.eurekalert.org/ release/colocolo 071999.htm/ (9/1/00) and http://www.sciencedaly.com/releases/1999/07/990722065100.html (10/11/00).

Barkman, Henrik O. (25 Aug. 1994). "Celsius, Andres-The Man with the Thermometer." Project Galactic Guide. http://www.energy.ca.gov/education/scientists/celsius.html (2/20/01).

Bjarnason, Agust B. (1999). "Carbon Dioxide, Is the Earth Warming? Things Are Not Always What They Seem." http://www.rt.is/hab/

sol/sol-e.htm (6/5/01).
Bjornerud, Marcia. (n.d.). "Gaia: Science, Metaphor, or Myth? The Tangled Web." Lawrence University. http://www.lawrence.edu/dept/environmental_studies/gaia.html (9/6/00).
Boyd, Robert S. (7 Aug. 1998). "Scientific Evidence Shows Earth Once Was a Big Snowball." Knight-Ridder Tribune News. http://www.chron.com/ (12/9/00).
Britt, Robert Roy. (12 July 1999). "Great Civilization on the Nile May Owe Its Start to Climate Change." Space.Com, Inc. http://explorezone.com/archives/99_07/12_sahara.htm (10/13/00).
Broecker, Wally. (19 June 1998). "The North Atlantic Flip-Flop Disaster." Institute of Geophysics and Planetary Physics, Earth and Environmental Sciences Division, Los Alamos National Laboratory, Los Alamos, New Mexico. http://www.igpp.lanl.gov/climate.html (9/5/00).
Bung, Hans-P Peter and Stephen P. Grand. (18 May 2000). "Simulated Influences of Lake Agassiz on the Climate of Center North America 11,000 Years Ago." Nature Asia, Vol. 495, p. 334. Nature Publishing Group 2000, England.
Bushing, William W. (1999). "Weather and Climate: Little Ice Age." Santa Catalina Island Conservancy, California. http://www.catlinaconservancy.org/weather/lia.htm (10/22/00).
"California's Environment Threatened by Global W Warming: Water Problems, Wildfires to Increase, Impacts on Habitats, Quality of Life." (14 Nov. 1999). Union of Concerned Scientists, Cambridge, Massachusetts. http://www.uscusa.org/globalresources/index.html (9/10/00).
Calvin, William H. (19 Jan.1998). "The Great Climate Flip-Flop" The Atlantic Monthly. http://www.theatlantic.com/ issues/98jan/climate.htm (9/4/00).
Carlisle, John. (2 April 2000). "Cooling Off On Global Warming." National Policy Analysis. The National Center for Public Policy Research. http://www.nationalcenter.org.NPA284.html (9/11/00).
Carlisle, John. (19 April 1998). "Global Warming: Enjoy It While You Can." National Policy Analysis. The National Center for Public Policy Research. http://www.nationalcenter.org (9/5/00).
Carter, Rae, Sean LeRoy, Trisalyn Nelson, Colin P. Laroque and Dan J. Smith. (1999). "Dendroglaciological Investigations at Hilda Creek

Rock Glacier, Banff National Park, Canadian Rocky Mountains." Department of Geography, University of Victoria, Victoria, Canada. http://office.geog.uvic.ca/dept/grad/students/colin/tree.htm (10/12/00).

Carver, Glenn, Owen Garrett, Hubert Teyssedre and Olaf Morgenstern. (19 Oct. 1999). "The Ozone Hole Tour." Centre for Atmospheric Science, Cambridge University, United Kingdom. http://geography.about.com/science/geography (9/2/00).

Christy, John R. and Roy W. Spencer. (2000). "Global Atmospheric Temperatures of the Lower Troposphere and Lower Stratosphere from the Microwave Sound Units." University of Alabama in Huntsville, Huntsville, Alabama. http://capita.wustl.edu (10/14/00).

"Climate Change and Greenhouse Gases" (December 1998). http://geology.about.com/msub31.htm (9/2/00).

"Climatologist: Global Warming Theory Based on Skewed Data." (26 Jan. 1999). Associated Press. http://www.junkscience.com/jan1999/skewed.htm (9/5/00).

Coffey, John. (19 April 2000). "Global Warming?" Atmospheric Scientists on Greenhouse Warming. http://www.xmission.com/ ~vote/gw.htm (9/4/00).

Corbyn, Piers. (19 Nov. 1997). "A (Solar-Based) New Alternative to Theories of Global Warming." Weather Action and South Bank University, South Bank Technopark, London, United Kingdom. http://www.millengroup.com/ (9/20/00).

Cutler, Alan. (13 Aug. 1997). "The Little Ice Age: When Global Cooling Gripped the World." The Washington Post. http://www.vehiclechoice.org/climate/cutler.html (9/2/00).

Daly, Andrew. (8 May 1998). "The Little Ice Age; Was it Big Enough to Be Global?" http://www.jrscience.wcp.muohio.edu/weather/PaperProposalArticles/ TheLittleIceAGEwasitbigeh/html (10/11/00).

Daly, John L. (29 July 2000). "Santa Takes a Mid-Summer Swim." Pine Lake, Tasmania, Australia. http://www.vision.net~daly/ (9/4/00).

Daly, John L. (2 Jun. 2000). "Testing the Waters, a Report on Sea Level." Greening Earth Society. http://www. Greeningearthsociety.org/Articles/2000/sea.htm (9/4/00).

Devine, Jim. (1999). "William Thomson Baron Kelvin of Largs (1824-1907)." Hunterian Museum, University of Glasgow. http://www.hunterian.gla.ac.uk/collections/science.html (2/20/01).

Dietze, Peter. (19 July 1999). "Estimation of the Solar Fraction and

Svensmark Factor." Climate Change: Guest Papers. http://www.microtech.com.au/daly/fraction/ Fraction/fraction.htm (6/5/01).

Dillin, John. (24 Aug. 2000). "Global Cooling—Mini-Ice Age" The Christian Science Monitor, The Christian Science Publishing Society. http://www.csmonitor.com/durable/2000/08/24/p1682.htm (10/13/00).

Dmitriev, Alexey N. (January 8, 1998). "Planetophysical State of the Earth and Life." IICA Transactions, Vol. 4., Siberian Department of Russian Academy of Sciences. The Millennium Group. http://www.millengroup.com/repository/global/ global.html (9/20/00).

Dougherty, Jon. (1997). "Global Cooling to Global Warming: The Alarmists Can't Decide." Covenant Syndicated, Vol. 1, No. 68. USA Features Media Co. http://capo.org/ (9/2/00).

"Dust Is Key to Global Warming, Studies Find." (2 April 1996). Reuter Information Service. http://www.nando.net/newsroom/ntn/world/040396/world_3116.html (9/5/00).

"Early Warning Signs of Global Warming." (12 Jun. 2000). Union of Concerned Scientists, Cambridge, Massachusetts. http://www.uscusa.org/globalresources/index.html (9/10/00).

Edwards, Rob. (27 Nov. 1999). "Freezing Future." New Scientist. http://www.newscientists/news/slime.html (9/5/00).

Elias, Scott. (1997). "Welcome to the World of Ice Age Paleoecology!" National Science Foundation, Boulder, Colorado. http://cutter.colorado.edu.1030/~saelias.html (915/00).

"Fahrenheit, Daniel Gabriel" (2000). Britannica.com, Inc. http://www.britannica.com/seo/d/daniel-gabriel-fahrenheit/ (2/20/01).

"Fahrenheit, Daniel Gabriel 1686-1736." (20 Feb. 2001). BBC Online. http://www.bcc.co.uk/history/programmes (2/20/01).

Farb, Michael. (1995). "James Lovelock's Gaia Hypothesis: Past, Present, and Future." http://www.slip.net/ (9/5/00).

"Final Word on Global Warming, A." (Aug. 2000). http://www.ngdc.noaa.gov/paleo/globalwarming/end.html (9/20/00).

"Food Outlook: Global Information and Early Warning System on Food and Agriculture." (June 2000). Food and Agriculture Organization of the United Nations, Commodities and Trade Division, Rome, Italy. http://www.fao.org (9/4/00).

"Forecasting Epidemics." (2 Oct. 2000). "New Eyes on the Ocean." National Geographic, Vol.198, No. 4, p. 111.

Fortner, Linda. (2001). "A Little History of Orchids." http://www.

orchidlady/history.html (3/2/01).
Friend, Tim. (9 Sept. 1997). "Team of Scientists Looks at New Phenomenon of Global Cooling." USA Today, National News. http://detnews.com/1997/nation/9709/10/09/09100010.htm (9/4/00).
"Gaia Hypothesis, The." (23 May 1999). http://www.kheper.auz.com/gaia/Gaia_Hypothesis.html (9/6/00).
"Gaia Hypothesis." (1999). The Remarkable Ocean World. http://www.oceansoline.com/gaiaho.html (9/5/00).
Gidwitz, Tom. (Mar./April 2000). "Telling Time." Archaeology. www.archaeology.org (3/17/01).
"Global Cooling." (25 May 1998). http://www.muntwashington.org/notebook/transcripts/1998/05/25.html (9/10/00).
"Global Cooling—A Possible Whitehouse Effect?" (2000). http://guernsey.uoregon.edu/ (9/12/00).
"Global Warming." (2000). Environmental Defense, Global and Regional Air Program. New York, New York. http://www.edf.org/programs/GRAP/ (9/20/00).
"Global Warming." (n.d.). http://www.epa.gov/ (9/4/00).
"Global Warming." (10 December 1999). National Oceanic and Atmospheric Administration. http://geology.about.com/ (9/2/00).
"Global Warming in Brief ." (n.d.). Global Warming Information Page. http://www.globalwarming.org/brochure.html (9/3/00).
"Global Warming and Climate Change." (1994). Carnegie Mellon University, Department of Engineering and Public Policy, Pittsburgh, Pennsylvania. http://www.gcrio.org/ (9/2/00).
"Global Warming Debate." (n.d.). Global Warming Information Page. www.globalwarming.org (9/4/00).
"Global Warming: Early Warning Signs." (1999). http://www.climatehotmap.org (9/10/00).
"Global Warming: Focus on the Future." (1997). EDF. http://www.enviroweb.org/edf/ (9/12/00).
"Goddess of the Earth? The Debate over the Gaia Hypothesis." (1997). University of Michigan. http://www.spnl.umich.edu/ (9/5/00).
Goffman, Joseph and Daniel Dudek. (19 April 1998). "Spurring Early Greenhouse Gas Reductions in the U.S." Environmental Defense, Global and Regional Air Program. Newsletter, Vol. XXIX, No. 2. New York, New York. http://www.edf.org/ (9/20/00).
Green, Kenneth. (1999). "Questions People Ask About Climate Change." Reason Public Policy Institute, Los Angeles, California. http://

www.reason.org/climatefaqs.html (9/4/00).

Greene, Arthur M., Wallace S. Broecker and David Rind. (1 July1999). "Swiss Glacier Recession Since the Little Ice Age: Reconciliation with Climate Records." Geophysical Research Letters, Vol. 26, No.13, pp. 1909-1912. http://www.agu.org/ (10/12/00).

Griesbach, R.J. (2000). "Potted Phalaenopsis Orchid Production: History Present Status, and Challenges for the Future." Floral and Nursery Plant Research, U.S. National Arboretum, Beltsville, Maryland. http://primera.tamu.edu/ orchids/griesbach.htm (3/2/01).

Hammersmark, Erick. (2001). "The Metric System." Grolier Electronic Publishing, Inc. http://www.uffda.com/~bink/ metric.html (3/2/01).

Hardy, Douglas R. and Raymond S. Bradley. (13 Dec. 1996). "Climate Variability in the Americas from High Elevation Ice Cores." Inter-American Institute for Global Change Research, Bariloche, Argentina. http://www.geo.umass.edu (10/11/00).

"Harvard-Smithsonian Physicists Calls Sun's Variability the Greatest Factor in Global Climate Change." (7 August 1998). The National Consumer Coalition's Climate Change Working Group. www.globalwarming.org (10/12/00).

Hauser, Rachel. (3 Mar. 2000). "Polar Paradox." Earth Observatory. http://geography.about.com/science/ geography/ (9/2/00).

Hieb, Monte and Harrison Hieb. (28 Nov 1999). "Global Warming: A Chilling Perspective." http://www.clearlight.com/ (10/13/00).

Hillger, Don. (16 Jan. 2001). "A Chronology of the SI Metric System." U.S. Metric Association (USMA), Inc. http://lamar.colostate.edu/ (3/1/01).

"Historical CO2 Records from the Law Dome DE08, DEE08-8-2, and DSS Ice Cores, DIAC/Trends." (5 Mar 2000). Neonet News. The Netherlands. http://www.neonet.nl/ (10/11/00).

Hodges, Glenn. (Mar. 2000). "The New Cold War: Stalking Arctic Climate Change by Submarine." National Geographic, Vol. 197, No. 3, pp. 30-41.

Hoffman, Paul F. and Daniel P. Schrag. (Jan. 2000). "Snowball Earth." Scientific American. http://www. sciam.com/ (10/16/00).

Hooker, Richard. (6 Jun. 2000). "Cultures in America." The People, World Civilizations. http://www.wsu/ (3/31/00).

Hosler, Charles R. (31 August 1998). "Cooling Off on Global Warming." The Chapel Hill (North Carolina) Herald. www. globalwarming.org

(9/4/00).

Hotz, Robert Lee. (13 Jan. 2000). "Panel: Global Warming Is a Reality." Los Angeles Times. http://seattletimes.nwsource.com/ news/nations-world/htm98 (9/10/00).

Hoyt, Douglas V. (14 Jan. 2000). "Greenhouse Warming: Fact, Hypothesis, or Myth?" http://users.erols.com/ (9/5/00).

"Ice Ages." (1997). Snowtastic Snow. http://tqinior.thinkquest.org/3876/iceage.html (10/13/00).

"Ice Ages." (1995). Illinois State Museum. http://www.museum.state.il.us/exhibits/ice_ages (9/2/00).

"Ice Ages, The: Are They Over?" (15 Sept. 1997). http://www. geography.about.com/science/geography/library/weekly/ (8/31/00).

"Ice Ages and Glaciations." (Jun. 2000). http://www.haartwick.edu/geology/ (9/3/00).

"Little Ice Age Holds Big Climate Clues." (1 Mar. 2000). Environmental News Network. http://www.cnn.com (10/13/00).

"Impact and Dinosaurs." (31 August 1997). http://kidsastronomy.about.com/kids/kidsastronomy/library/weeky (9/3/00).

Johnson, Robert G. (8 July 1997). "A Modest Proposal." EOS. http://geology.about.com/science/geology/ (8/31/00).

Jones, Philip. (19 June 1999). "Global Temperatures." The Climate Research Unit, United Kingdom. http://www.cru.uea.uk/cru/annrep 93/globtemp.htm (9/5/00).

Karl, Thomas R. (1997). "Recent Rise of Night Time Temperatures." National Climatic Data Center, Asheville, North Carolina. http://captia.wustl.edu/NEW/karl.html (10/13/00).

Karl, Thomas, R., and Nathaniel B. Guttman. (2000). "Global Warming Update." National Climatic Data Center, Asheville, North Carolina. http://captia.wustl.edu/NEW/karl_up.html (10/14/00).

Karl, Thomas R., Neville Nicholls and J Jonathan Gregory. (19 May 1997). "The Coming Climate." Scientific American. http://www.sciam.com/0597issue/0597Karl.html (9/4/00).

Keller, C.F.F. (19 June 1998). "Global Warming: An Update." Institute of Geophysics and Planetary Physics, Earth and Environmental Sciences Division, Los Alamos National Laboratory, Los Alamos, New Mexico. http://www.igpp.lanl.gov/ (9/5/00).

"Kelvin, William Thomson, 1st Baron" (2001). Funk and Wagnalls.com. http://www.fwkc.com/encyclopedia/low/ articles/ (2/20/01).

Lassen, K. (1998). "Long-Term Variations in Solar Activity and Their

Apparent Effect on the Earth's Climate." Danish Meteorological Institute, Solar-Terrestrial Physics Division, Copenhagen, Denmark. http://www.millengroup.com/ (9/20/00).

"Last Glacial Epoch, The." (2000). http://library.thinkquest.org (10/13/00).

Ledley, Tamara S., Eric T. Sundquist, Stephen E. Schwartz, Dorothy K. Hall, Jack D. Fellows, and Timothy L. Killeen. (28 Sept. 1999). "Climate Change and Greenhouse Gases" EOS, Vol. 80, No. 39, p. 453. http://geology.about.com/ msub31.htm (9/6/00).

Lemonic, Michael D. (31 Jan. 1994). "The Ice Age Cometh? Last Week's Big Chill Was a Reminder that the Earth's Climate Can Change at Any Time." Time Domestic, Time, Inc. http://www.time.com/time/magazine/archive/ (10/13/00).

Leifert, Harvey. (26 Oct. 1998). "Long Lava Flows May Have Taken Years, Causing Global Cooling and Extinctions." American Geophysical Union, Washington, D.C. http://www/agu.org/sci_soc/prrl/prrl9834.html (8/31/00).

Lioubimtseva, E.U. (1), Gorshkov S. P. (2) and Adams, J.M. (3). (1999). "A Giant Siberian Lake During the Last Glacial: Evidence and Implications." (1, 2) Department of Physical Geography, Faculty of Geography, Moscow State University, Moscow, Russia, and (3) Environment Sciences Division, Oak Ridge National Laboratory, Oak Ridge, Tennessee. http://www.esd.ornl.gov/projects/ (9/3/00).

Lovelock, James. (1996). "What Is Gaia?" http://ori.com/ (9/10/00).

Lovelock, James and Lynn Margulis. (1996). The Gaia Hypothesis. Mountain Man Graphics, Australia. http://www.magna.com.au/~prfbrown/gaia-jim.html (9/5/00).

Loy, Jim. (2000). "The Little Ice Age." http://www.mcn.net/~jimloy/little.html (10/12/00).

MacCracken, Michael C. (24 Aug. 1995). "Climate Change: The Evidence Mounts Up." Nature, vol. 376, pp. 645-646.

Marcus, Adam. (12 Dec. 1997). "Model Shows Certain Gasses Could Stimulate Global Cooling." University of Michigan. http://www.eurekalert.org/releases/am-msegsge.htm/ (9/1/00).

"Mass Extinctions: Introduction." (2000). BCC Evolution Weekend. http://www.bbc.co.uk/education/darwin/exfiles/massintro (2/3/01).

Marquis, Colin and Stu Ostro. (29 Aug. 1999). "Is the Weather Getting Worse?" USA Weekend. http://www.usaweekend.com (9/5/00).

Mayewski, Paul A. and Michael Bender. (1995). "The GISP2 Ice Core Record-Paleoclimate Highlights." U.S. National Report to IUGG, 1991-1994. Rev. Geophys. Vol. 33 Suppl., American Geophysical Union. http://earth.agu.org/revgeophys/ mayews01/mayews01.html (10/13/00).

Mayewski, Paul A., Johon Bolzan, Debra, Meese, Todd Sowers, Mark Twickler, and Gregory Zielinski. (May 1998). "Ice Core Contributions to Global Change Research: Past Successes and Future Directions." Ice Core Working Group, Office of Polar Programs, National Science Foundation. http://www.nici-smo.sr.unh.edu/icwghtml.html 10/11/00).

McKibben, Bill. (1997). "The Earth Does a Slow Burn" New York Times, New York, New York. http://www.edf.org (9/20/00).

"Mean-ingless Measures?" (1998). http://www.stats.org (9/5/00).

"Metric System." (2000). Encarta Online Encyclopedia. http://encarta.msn.com/index/conciseindex/26/02694000.htm (3/1/01).

Meyer, Peter. (4 Jan. 2001). "The Julian and Gregorian Calendars." http://serendipity.magnet.ch/hermetic/ cal_stud/cal_art.htm (2/5/01).

Michaels, Patrick J. (31 Dec. 1998). "Past Climates." The Cato Institute. http://www.cato.org/pubs/pas/pa-329es.thml (9/4/00).

Miller, Stephen. (1989)."Gaia Hypothesis." http://erg.ucd.ie/arupa/references/gaia.html (9/5/00).

Monastersky, Richard. (April 1996). "Health in the Hot Zone: How Would Global Warming Affect Humans?" Science News, Science Service, Inc. http://dieoff.org/page70.htm (2/25/01).

Morrison, Roy. (1995). "An Example of the Catastrophic View: A Global Warming Scenario." Ecological Democracy, South End Press. http://dieoff.org/page26.htm (2/24/01).

"Mount Pinatub Pinatubo."(n.d.). http://ww2.sunystuffolk.edu/mandias/ (8/31/00).

"Mount St. Helens." (n.d.). http://ww2.sunystuffolk.edu/mandias/ honors.student/volcano (8/31/00).

"National Climate Assessment: A Wake —Up Call." (12 Jun. 2000). Union of Concerned Scientists, Cambridge, Massachusetts. http://www.uscusa.org/globalresources/ index.html (9/10/00).

"North American Drought: A Paleo Perspective." (31 July 2000). NOAA Paleoclimatology Program. http://www.ngdc.noaa.gov/ (9/3/00).

"On the Shoulders of Giants: Milutin Milankovitch (1879-1958)." (n.d.). Earth Observatory. http://geography.about.com/science/ (9/2/00).

Olsen, Paul. (1997). "Earth's First 3.7 Billion Years." Columbia University, New York, New York. http://rainbow.Ideocolumbia.edu/courses/v1001/7.htm (2/3/01).

"Paleo Perspective on Global Warming, A" (Aug. 2000). NOAA. http://www.ngdc.noaa.gov/paleo/global/warming/ (9/12/00).

"Paleoclimatologist Sees Another Ice Age Coming." (6 Feb. 2000). UniSci: Daily University Science News. http://unisci.com/stories/20001/02/6001.html (9/10/00).

Pearce, Fred. (19 July 1997). "They're Among the World's Top Scientists and They Don't Believe in Global Warming." New Scientist. http://geology.about.com/msub31.htm (9/2/00).

Poling, Jeff. (2 Aug. 1999). "Global Warming Can Cause Global Cooling." http://www.dinosauria.com/ (8/31/00).

"Population Growth Could Affect Global Warming." (2000). Environmental Defense, Global and Regional Air Program. Newsletter, Vol. XXVII, No. 6. New York, New York. http://www.edf.org/programs/GRAP/ (9/20/00).

Price, David. (4 Mar. 1995). "Energy and Human Evolution." Population and Environment: A Journal of Interdisciplinary Studies, Vol. 16, No. 4, pp. 301-19. http://dieoff.org/ page137.htm (2/24/01).

"Quick Background to the Last Ice Age, A." (Jun. 1999). http://www.esd.ornl.gov/projects/qen/nerc.html (9/3/00).

Racer, Paul. (13 Oct. 1997). "Scientists Believe Little Ice Age May Blunt Effect of Earth's Global Warming." Associated Press, Athens, Newspapers, Inc. http://www. athennewspapers.com/ (10/13/00).

Rayburn, John A. (5 Nov. 1996). "Modeling Isostatic Rebound to Evaluate Lake Agassiz's Role in the Younger Dryas Global Cooling Episode." Department of Geological Sciences, University of Manitoba, Canada. http://www.cc.umanitoba.ca/ (10/11/00).

Reagan, Michael. (2000). "Should We Worry About Global Warming or Global Cooling Trends? Global Warming or Global Cooling the Threat for the Future?" The Reagan Information Interchange. http://www.reagan.com/ (9/12/00).

Redford, Gabrielle deGroot. (April/ May 2001). "Why Should We Care about Marine Worms?" National Wildlife, Vol. 39, No. 3, p. 12. www.nwf.org (10 April 2001).

Roach, John. (1 Mar. 2000). "Little Ice Age Holds Big Climate Clues." Environmental News Network. http://www.enn.com/ (10/12/00).

Robinson, Arthur. (21 Aug. 2000). "Global Liars." The New Australian, Melbourne, Victoria, Australia, No. 16. http://www.newaus.com.au/indiex.html 9/4/00).
Rozell, Ned. ed. (24 Dec. 1996). "Daniel Fahrenheit, Andres Celsius Left Their Marks." Alaska Science Forum. http://www.gi.alaska.edu (2/20/01).
"Secrets of the Ice" (2000). http://www.secretoftheice.org/ (10/13/00).
"Some Comments on the Manic Sun." (2000). http://www.euronet.nl/users/e_wesker/solspot.html (10/18/00).
Spencer, Roy. (6 Feb. 1997). "Is Temperature Up or Down or Both?" Marshall Space Flight Center, Huntsville, Alabama. http://science.msfc.nasa.gov/newhome/headlines/essd5feb97.1.html (9/2/00).
Spencer, Roy. (August 1998). "Measuring the Temperature of Earth from Space." Global Warming Information Page. www.globalwarming.org (9/4/00).
"Statement by Atmospheric Scientists on Greenhouse Warming." (19 April 2000). http://www.xmission.com/ ~vote/gw.htm (9/4/00).
"Story of Orchids." (N.D.) http://www.orchid.org.uk/contents.htm (3/2/01).
Sunspots and the Solar Cycle." (24 Jan. 1998). Bishop Web Works. http://www.sunspotcycle.com/ (10/12/00).
Svensmark, Henrik and Eigil Friis-Christensen. (1997). "Variation of Cosmic Ray Flux and Global Cloud Coverage—a Missing Link in Solar-Climate Relationship." Journal of Atmospheric and Solar-Terrestrial Physics, 59 (11), 1225-1232. http://www.dsri.dk/~hsv/ (6/7/01).
"Tambora." (n.d.). http://ww2.sunystuffolk.edu/mandias/ (8/31/00).
Taylor, George H. (Sept. 1999). "Blaming Disasters on Global Warming Doesn't Help." American Association of State Climatologists. http://sepp.org/NewsSEPP/GeorgeTaylor.html (9/5/00).
Thurlow, Dave. (25 May 1998). "Global Cooling." The Weather Notebook. http://www.mountwashington.org/ (9/15/00).
Topfer, Klaus. (1999). "Climate Information Kit." United Nations Environment Programme's Information Unit for Conventions. Geneva, Switzerland. http://geography.about.com/science/ (9/4/00).
Trenberth, Kevin E. (1997). "The Use and Abuse of Climate Models." Nature 386: 131-133.
Tyson, Peter. (2000). "Stories in the Ice." NOVA Online, WGBH T.V. http://www.pbs.org/wgbh/nova/warnings/ (10/12/00).

Van Helden, Albert. (1995). "The Sun." Rice University. http://seds.lpl.arizona.edu/nineplanets/sol.html (10/12/00).

"Volcanic Effects on El Nino." (n.d.). http://ww2.sunystuffolk.edu/mandias/ (8/31/00).

"Volcanic Effects on Global Cooling." (n.d.). http://ww2.sunysffolk.edu/mandias/ (8/31/00).

"Volcanic Effects on Ozone Depletion." (n.d.). http://ww2.sunysffolk.edu/mandias/ (8/31/00).

"Volcanoes and Global Cooling." (1996). Department of Meteorology, University of Maryland College Park. http://www.meto.edu/~owen/chpi/images/volceff.html (8/31/00).

Walker, Gabrielle. (6 Nov. 1999). "Snowball Earth." New Scientist. http://www.newscientist.com/ 12/9/00).

"Warming of the Earth." (1999). The Woods Hole Research Center. http://geography.about.com/science/geography/ (9/2/00).

Wesker, Evert. (16 July 2000). "Climate Change: A Summary of Some Present Knowledge and Theories." http://www.euronet.nl/users/_wesker/climate.html (10/12/00).

Wesker, Evert. (8 Aug. 2000). "El Nino and Global Warming." http://www.uronet.nl/users/e_wesker/e/ninow.html (10/18/00).

"Why Are We So Afraid of the Sun?" (1999). http://www.greatdreams.com/sun.html (10/11/00).

Witze, Alexandra. (23 Nov. 1998). "Theory Says Climate Change Depends on Solar Wind/Cosmic Rays." Dallas Morning News, Dallas, Texas. http://www.millengroup.com/ (9/20/00).

"World Study Blames Global Warming for Miles + of Arctic Ice Melting Yearly." (3 Dec. 1999). Tampa Tribune, Tampa Florida. http://www.global-warming.com/ (9/10/00).

"Wyoming Ice Hints at Abrupt End to Little Ice Age." (29 Feb. 2000). explorezone.com.http://explorezone.com/ (10/13/00).

Zipperer, Rich. (2000). "A Climate Change Glossary." National Consumer Coalition. http://www.globalwarming.org/ glossary.htm (10/13/00).

Printed in the United States
874800003B